The Science Behind

FREE WILL

By

Timothy Michaels

Table of Contents

Preface

I began thinking about free will and putting together the ideas for this book after discovering the physical aspects of our world which I believe are key to making it work.

My primary interest in the topic has been physics, but my faith has grown through that pursuit, as it has through all of my other scientific study. This came about because I had to develop new interpretations of existing information and theories, so that they could make sense to me. What I found is that the evidence all fits together. While developing this new way of looking at physics I saw how it fits with religious teachings. And for me, this unified religion and science.

Looking at free will, I saw how much we seem to be like the creator of the world that we implement our will similar to the way he does. I also saw the significance of our freedom to any master plan there may be. Any purpose to creation seems to rely on this freedom.

Certainly, in looking at our world with its great size and complexity, there are many questions to ask. As I explore, I try to remain focused on one question at a time and remember that everything I learn about each topic must fit with the others, since they are all ultimately connected.

I don't believe there are real contradictions in our world, but rather that each one is actually a puzzle (paradox) yet to be solved. The great physicist and colleague of Einstein, Neils Bohr, once said, "How wonderful that we have met with paradox. Now we have some hope of making progress."

My explanations rely heavily on the most established principles we know of in order to provide readers with reasoning for how my own ideas are also reliable. With each idea that I introduce, I walk readers through the logic behind it so that they may decide for themselves how much weight to give it. At the same time, I don't want to go into excessive scientific detail about concepts. Instead, I provide explanations of what they need to know, as well as enough information to allow readers to pursue further knowledge in those areas if they choose.

One of the concepts I cover in this book is my own theory of how movement occurs at the subatomic, or most fundamental, level. This subject is explained in detail in my book *Out of This World: The Movement Dimension*. Movement is important to the execution of free will. It's the final phase of the process. This theory describes a link, or interface, between physical and non-physical dimensions and entities. It explains how a ghost may grab a candlestick.

In my book, *Absolute Relativity: How Newton and Einstein Agree*, I explain the version of relativity that was applied in movement theory. It provides scientific evidence to support movement theory and it's non-physical dimension which I refer to in this book. The two theories rely on each other. Absolute relativity also establishes that there is only one reality which we each have our own experience of. But in this book we will be focusing on movement theory's role in free will.

Time is another subject in which I describe a structure that differs from the way many of today's physicists understand it. Instead, I explain and apply time concepts in a way that I believe makes the most sense.

I always try to normalize scientific ideas and make them as simple and understandable as they can be. But most importantly, they must be logical and fit together in ways which are not strange, but instead are believable.

While I do use new and unique scientific ideas to explain how free will works, I believe the most discerning and logically thinking readers will find these ideas to be sound and in agreement with experience, evidence, and our most reliable scientific theories.

I also provide new interpretations of experimental data and real-life experience which fit into a whole view of our world, while solving many of the paradoxes given to us by today's more popular interpretations. Overall, what I present is a hypothesis. My intent is to show what the truth may be, so that readers may decide for themselves what they believe.

Chapter 1 - Introduction

Scientists like to dismantle nature in order to discover more of it. They take apart objects to find what they're made of. Then take apart those objects too. They've discovered that people are made of cells, which are made of molecules, which are made of atoms, which are made of electrons, protons, and neutrons. And protons and neutrons are made of quarks held together by gluons. Some say those are all made of strings.

I say they're made of a fundamental substance which may be most comparable to the classic concept of the ether. Regardless of what it is, we describe their behavior with quantum physics.

It seems we've gotten to, or near to, the bottom of things. Even if there's more to take apart, we at least know a lot about the behavior at the subatomic level. And so far, we have not found free will, nor have we found a mind. We haven't found how humans have an ability to want or choose, yet somehow we do.

I'm going to lead you down a path of exploration and show you where we are in this search. Then I'm going to show you a way of assembling what we know along with some of my own ideas which may resolve the issue. If you believe this is correct, then you may find yourself with a heightened appreciation for life. It may also give you a greater sense of responsibility for reaching your destination as an individual and ours as humanity.

First we need to know what it is we're looking for so we can recognize it when we see it.

Free will has two parts, freedom and will. Will is our wanting. It's the "us" part of it.

Will

Little seems to be known of our wanting or our ability to want.

Still, it seems to be a fundamental part of us and is unique to each of us. You likely consider your will as very personal and special, your most prized possession to preserve through all difficulty. You probably also consider it as something no one can take from you, perhaps because it can't be removed or maybe because it's just worth so much to you that you would protect it at any cost.

As valuable as it is to us, many scientists claim it's an illusion. They say that deep inside you, that thing you guard the most isn't real. And if someone could somehow unlock your most private safe they would find nothing at all.

Freedom

The second part and the part which I intend to focus on is freedom. I believe it is most accessible to our understanding and it may unlock a greater level of faith for many readers.

Freedom is the ability this world gives us to make choices. Since we are physical beings, we can search physical laws for this ability, and by now we know a lot about physical laws. It may be true that what we know is only a small portion of what there is to learn. However, we know some very important concepts and I believe that those are enough to guide us to an answer of how it is we have the freedom to exercise our will. Along the way you may also see that your will, whether we understand it or not, is real.

We'll be discussing our own will and our ability to implement it. And we'll also explore the will of our creator since I believe they are closely related. This has a lot of religious implications since it relates to the creator, and also because I suggest that our will is not physical. So we'll cross back and forth between concepts of the physical world and the logic which informs us of the non-physical.

This type of investigation puts us right into religious territory. While it's not my objective to explain religious ideas, they are unavoidable. And just as unavoidable is the obvious tension between science and religion. So we'll begin by looking at this relationship.

Chapter 2 – Science vs Religion

All religions, with the exception of materialism, have had increasing difficulty when confronted by the success of science. This is because most religions are based on there being a non-physical reality. At a minimum, they involve something or someone which is not physical, but is outside of the dimensions we live in. Materialism avoids this by considering only what exists in space and time.

The Influence of Science

As science advances at a quickening rate, it explains more and more of our world using only physical reasoning. This is what science has become, logic of the physical.

While science originated from the Latin word for knowledge, it is considered to be limited to knowledge regarding observed facts which have been repeatedly tested and found true. It seems to have evolved to refer only to the physical.

And as it advances, it achieves greater precision as well. It is on such a roll that it would be easy to imagine that science may someday be able to explain everything; that it may be able to answer every question one could ask. Not only is it providing reasons, but it's achieving precision in them, allowing no wiggle room for any non-physical involvement. And apparently, no room for a religious viewpoint.

Science has been so successful that it's created a circle of reasoning which excludes all that is not physical. It has achieved such a high level of certainty that we now defer to science for truth even though we know scientific concepts are always subject to advancement or being replaced by a newer, better theory. If one makes a statement about the non-physical now, it may be refuted as being non-scientific. And many people believe that if it's not scientific, it's not likely true, or is unreliable.

By avoiding a few questions, a person could live today believing that science explains everything.

Throughout modern history, religion has been losing ground to science. We've transitioned from religion explaining everything, to religion often only being invoked to explain why the world is the way it is, or perhaps just to point out who started the big bang.

A popular belief today seems to be that God wrote the physical laws then set off the big bang and let it go on its own. Those who explore deeper may conclude that he must have also guided evolution somehow. Many people believe he's not involved in our lives today. And those who believe he is, likely can't explain how it's possible and may feel forced to believe that if miracles occur, they must violate the laws of physics.

Mysticism

At the time when religion explained all, it could be said that the world operated by mysticism. Everything which happened was a mystery. Mysticism is essentially manipulation of the material world by a non-physical entity. It is something spiritual reaching into our world and causing things to happen.

So more specifically, it's mysticism which gave way to science. Not that the non-physical can't exist, but that science seems to exclude it from being involved in our world. The explanation for any occurrence can be categorized as either mysticism (non-physical influence) or natural law (science or physical influence). This is why materialism as a religion isn't suffering. It is a belief that only the physical is real, so that all rewards to be had, all purpose in life, is physical.

Today, as I talk to people of faith who do believe in a non-physical being, most are reluctant to admit a belief in mysticism or any physical manipulation of our world by anything non-physical. I suspect they don't want to give the appearance of disagreeing with science.

Some explain that God can change their luck or their chances. This is a reliance on all of the complexities of the world to cover up mysticism. Luck, or chance, is usually used to describe a situation in which the outcome is calculable but far too complex for our abilities to determine. A good example would be guessing when a car will break down. To change something like that seems that it must go against physical laws since we know that mechanical failure happens according to those laws.

When going further in conversation with the same people I usually find that they do understand that science explains all these things. Yet luck, or chance, are just the terms used to describe their inability to calculate their future. Any alternative requires an awkward admission of not having an answer, or a rational explanation, for their belief in God.

Many of us prefer to close our eyes to the conflict between religion and science, and it's understandable. We believe that both must be correct, but it's hard to see how that can be.

However, even faced with this apparent conflict I think we can open our eyes. I think we can accept the great success of science and maintain faith in a non-physical being who is actively involved in our world. We can do this with the understanding of how natural laws were constructed to direct all things physical, be precise, and still allow for all of the physical manipulation our religion tells us is real. This manipulation is by the involvement of a will greater than ours. Additionally, even our will is mystical if it's outside of the physical.

I believe our world was made perfect and we were given the freedom to do as we choose. Without freedom I don't see how we could fulfill any purpose we have. And how could there be a world without purpose?

To see how we could possibly be free we need to investigate the world we live in. First let's look at how we investigate.

Chapter 3 - Investigative Method

After so much advancement in science, some people seem to have changed their method of evaluating reality. They now believe we need to dig down into a subject in order to find reality or truth. But I believe that thought process is flawed. It can sometimes suggest that what's obvious in our world isn't real.

If truth is only found at the hidden fundamental levels of our world, then one might believe that what we perceive is an illusion. The flaw is in holding an underlying concept of something we can't see as being of higher regard than that which we do. That the concept of a greater truth somehow diminishes or reduces the truth of what's in front of us. I'm not suggesting there's nothing beyond perception or that something else isn't real, only that perceptions must be considered first.

Perceptions

Is what we perceive real? Asking this question seems that it may open the door to an endless series of similar questions and answers about science. But the solution is provided to us by another source, and that is knowledge that's been given to us, not endless layers of physical reality. We have faith that our senses aren't deceiving us.

Through scientific investigation we know that all of our senses are subject to defect and failure. We've all experienced seeing, hearing, smelling, tasting, or feeling something which was difficult to explain. Perhaps pondering a memory fooled us into believing we smelled our grandmother's cooking. Maybe we've felt a touch when nothing was there. This goes even beyond our real body so that about 60% of people who have lost body parts experience sensations that their limb is still there.

Our brain is constantly generating false sensations and it also has to decide which are from outside stimuli and which are internal. It isn't always correct. These hallucinations can come through into our consciousness especially when our senses are deprived, because our mind is seeking input.

All stimuli from our senses are brought to our brain by neurons and we know neurons are subject to failure by misfiring, leaking, or dying. Douglas Fox said in Scientific American that "the proteins that neurons use to generate electrical pulses, called ion channels, are inherently unstable."

We're in a constant struggle to remain in contact with reality. The mind is always checking and double-checking perceptions against each other and the behavior of objects against their expected behaviors.

Our connection to reality is by faith. And that stops the endless doubt. We proceed with confidence in this world by trust in something greater than us. That is that there is a greater truth than what our senses perceive. But that's not something we've searched for and found. It's a gift. As far as our searching is concerned, we have to work through our perceptions.

From faith in our perceptions, we're able to learn about the underlying makeup of our world by exploring the tangible which is in front of us. Certainly, any concepts derived from our firsthand experience cannot be any truer than the perceptions we base them on. The starting place in any investigation is with what we perceive and determine to be reliable in our world. If our perceptions aren't correct, then how can anything we derive or interpret from them be correct?

Simply, what's in front of us must be regarded as real in order for anything we learn from it to be real. An investigation must begin with what is known.

Seeing Is Believing

When Galileo Galilei (1564-1642) began using a telescope to look into space, he found that objects weren't exactly as they appeared using only his eyes. Mars had craters and mountains, and the sun had spots. Many people didn't believe him, even after looking through it for themselves. They suggested that it was a trick, that the telescope was somehow superimposing features on objects.

It took an investigation by 4 leading mathematicians in the year 1611 to come to the conclusion that what the unaided eye couldn't see is real.

This, I believe, is a good example of proper discernment of truth. We must begin by looking at the most obvious because it is what we can be most sure of and we should be critical of any interpretation of it.

So, I search the physical world to learn more about what's behind it. I look at proven theories and especially experimental evidence because the evidence is most real. Less certain is our interpretation of evidence.

When it comes to will, everything in my world indicates that I do have one. I do want things. And whatever my will may be a result of can't be any more real than my will itself.

Many speculations have been made about will having to arise from physics. Some claim that our will is an illusion created by our mind's need to rationalize every situation, even while it's short of information. No evidence has been found of our will being a phenomenon of the physical world. No one has been able to explain any specific mechanisms which wanting, our ability to decide for ourselves regardless of predictions, may arise from.

So we'll have to identify free will carefully, as a non-physical cause to a physical event. We're looking for the ability for things to happen in a way which has no physical explanation. This puts us in a difficult position as we'll now look at how precise the laws of nature are.

While many of us believe in free will already, by both faith and experience, it may be brought into question by anyone who explores science and doesn't see room for it. So we'll proceed down the path to that difficulty. Then we'll keep going to see what we find when we go further.

Chapter 4 - Our Predictable World

Throughout most of history it was easy for man to view himself as having free will. He wanted and he could choose. It was as simple as that. Meanwhile, he was learning that his world is knowable and exploring more of it, and also learning how to use it to his advantage.

Most people likely didn't think that there may be a conflict between the world being knowable and their ability to choose. But today we can see that the ideas seem to be fundamentally opposed to each other.

"I know the world and I can do what I want with it," seems to be a contradictory statement. The reason is because I am of the world. If the world is knowable and I am of it, then what I will do can be known, and I have no freedom to do as I want. The problem here is that I am physical and therefore bound by the laws of nature which make the world knowable. Knowing the world means it's predictable. And if the world is predictable, so am I.

However, my experience of the world is that it is extremely predictable in some ways and not so much in others. If one plant is good to eat, then every other plant which is like it is also good to eat. If stepping on one thorn hurts, then stepping on any of them will also hurt. But the weather and actions of people and animals are only somewhat predictable. And those are part of nature too. What makes them different is that their behavior is complicated. They seem to have a will of their own, even though we know that they too, are bound by laws of nature.

Their unpredictability seems to arise from our lack of knowledge of all the factors involved in their behavior. Scientists have found that they may not be unpredictable, we may just have an inability to predict them.

I know that a tree will grow toward the sun. Its behavior is easy to guess. But I don't know if a squirrel will go up the tree or across the ground. Its decision-making process is too complicated for me to predict its actions, in part because I don't know all of the factors it's considering. I could simply say that it has a will of its own. Or I could recognize its complexity. I don't know where the squirrel came from or if it has a destination already. I don't know what's on the ground or in the tree which may interest it. And I don't know if it's hungry. If I knew all of those things, perhaps I could predict what it would do. Or maybe not, if it does have free will. Complexity can be deceiving.

Complexity Or Will

In exploring science we gather and analyze all the factors we can in order to make predictions. A scientist is likely to optimistically claim that we will someday be able to predict the actions of a squirrel. While today we can't recreate its decision-making process, we can guess that its brain is so extremely complicated and the available input to it is so great that it could be calculating every move and we would not be able to tell.

The weather is an easier problem to evaluate. We don't see the sky as having a will like we have. By now we know enough about the brain to identify it as a place of decision-making. The sky has no brain, so how could it think? It's not even confined to a place. It's nearly everywhere. So, it's not a single object in the way a squirrel is. While a squirrel is obviously not human, it does have the same qualities we have which liken it to a thinking being. It has one consolidated body and a nervous system with a brain. The sky doesn't have any of that. It's just a lot of particles with no obvious order. The great number of particles with lots of room to move over a great distance gives it a complexity which we don't have the ability to predict. Only complexity makes the weather unpredictable. Perhaps only complexity makes the squirrel unpredictable too, and that complexity is inside its brain.

But I believe I'm different. I believe I'm not just complex but can actually make my own decisions. I want things and I can act accordingly. I don't climb a tree like a squirrel does because there's food up there. I climb it because I want to. That's the difference I see.

While calculating can become so complicated that we're unable to predict the results of a situation, it's not the same as thinking. Many scientists believe that thinking is just a phenomenon which results from great complexity. But so far, in making machines more complicated, we haven't been able to make them think.

Consider the mathematical powers and speed of a simple computer compared to your own mind. We can't calculate like a computer. And a computer can't think like us. Calculating, for us, is a learned process whereas creativity is natural. If I asked you to create a game you could do it very quickly. You could invent the rules faster than you could write them down. But when asked to determine whether the game makes sense and the rules all work together, it would take you much longer. Determining whether the game works is calculating because it's following logic. We can see by examples such as this that our minds are not calculators or computers. They're better at creating than following logic. If thinking was logic then math would be easy.

As scientists cut apart, separate, and analyze every part of the human body as well as animals' bodies, they find we're made of the same stuff which behaves in the same predictable ways. The stuff which the squirrel is made of, which earth and sky are made of, and which you and I are made of, are the same 92 chemical elements. (Actually 98% of our body is just a few of them, carbon, hydrogen, oxygen, and nitrogen, and a small amount of sulfur and phosphorous.) And all 92 elements are composed of only protons, neutrons, and electrons behaving according to the laws of physics.

Scientists find nothing which seems to be a free will. In fact, they don't find any method of computing or even any method of making decisions. In other words, no program. They only find connections, communication devices, neurons. We don't know how the brain thinks, only that different processes are associated with different areas of it. And if those areas become damaged, then other parts of the brain may take over those same functions. So maybe there is a non-physical aspect to us where decisions are made, or influenced. If so, that needs to interact with our physical body somehow.

Regarding our ability to act on our wants, perhaps nature is somehow just not precise. Maybe there's some variability allowing something as complex as the human brain to be unpredictable. The stuff which makes up our brain may somehow not be subject to strict physical laws. That would mean protons, neutrons, and electrons have a sort of freedom.

Perhaps there's a lack of precision to nature which gives us freedom so that we can mostly know our world, and still execute our will. (This would have to apply to all of physics since everything is made of these same fundamental parts.) Afterall, I do act on what I want.

We've identified two issues. One is that if there is a non-physical aspect to us, it needs a way to interact with our physical body. The other is that we need our body to have the freedom to move in a way which is not predictable by any method. We need to allow for non-physical causes to physical events. Both seem to go against science.

Chapter 5 - Precision of Nature

When man first realized that he could know things about the world, meaning he could make predictions, how could he have thought that it was precise, with all of its complexity? Everything he might predict could sometimes be wrong. Meat might be good to eat, or not if it's diseased or rotten. Fruit left on a tree may disappear overnight, taken by animals. Even the sun doesn't come up on a cloudy day. There were always exceptions.

If we watch objects fall enough times, we learn that things fall down, but our certainty is only as great as the number of times we see it occur. After we see it several times, we may decide we know what will happen, but only to that degree of certainty. Never absolutely.

In the past, knowing the world only meant being sure enough to make useful predictions. Absolute certainty and precision? Those concepts didn't exist among the general population. Some philosophers speculated that there may be an absoluteness to our world, once we remove all the exceptions which get in the way, but it could not have been more than a thought until about 400 years ago.

The first precise law of nature was written around 1609 by Galileo Galilei (1564-1642). He said distance is equal to 1/2 acceleration multiplied by time squared.

$$s = \frac{1}{2}at^2$$

And he was right. Not mostly right, or right most of the time. He was right all the time. (This formula is still used today.) We found that physical laws are precise.

Since then, scientists have discovered thousands of physical laws and been able to describe them with precise mathematical formulas such as Galileo's. Measurements have been refined, precision in our testing has increased, and today most deviation from predicted results of our best theories can be attributed to human error or our own inability to perform an experiment perfectly.

While not everything has been figured out, we have learned that physical laws seem to be precise. We may not yet have perfect formulas written, theories fully developed, or all phenomena accounted for, but we have good reason to believe that nature works precisely a certain way. Just what that way is in all situations is what we pursue now.

The state of our natural philosophy, or underlying beliefs about science, around the year 1800 was very well summarized by Pierre-Simon de Laplace (1749-1827) when he said, "We ought then to regard the present state of the universe as the effect of its anterior state and as the cause of the one which is to follow. Given for one instant an intelligence which could comprehend all the forces by which nature is animated and the respective situation of the beings who compose it - an intelligence sufficiently vast to submit these data to analysis - it would embrace in the same formula the movements of the greatest bodies of the universe and those of the lightest atom; for it, nothing would be uncertain and the future, as the past, would be present to its eyes."

This describes our whole universe as a very complex machine in which every detail of its operation is predetermined, the exact moment every leaf falls, the landing of every raindrop, and even every action and every thought you or anyone else will ever have.

This is a tremendous milestone of learning. In just 200 years we went from our first finding of precision to recognizing the absolute perfection of the whole world.

While today we know that LaPlace's statement isn't entirely correct, the future can't be predicted by any intelligence, his underlying realization of a perfectly logical world, I believe, is correct. The reason has to do with another foundational concept we learned about in the 1900s.

Chapter 6 - Quantum Probability

If you're familiar with the quantum effects, you may have expected this topic. As I concluded that physical laws are precise, I appeared to have left out quantum physics, but I haven't. It too is precise, so far as we have found. And there's no reason to believe otherwise.

Quantum probability is the concept that everything in our world does not have definite properties but instead has most probable properties. This effect is commonly relegated to only the smallest most fundamental particles, but actually applies to everything.

Those who believe in free will and are familiar with quantum physics commonly attribute our ability to implement our will in a precise world to the unpredictability of quantum effects. They see probabilities as a tool for freedom because it makes all predictions imprecise. And this seems to mean we can get away with a little bit of interference in the mechanical operation of our world. This appears to be the opening in natural law to allow for tinkering. But I don't believe a conclusion can be reached so simply. We need to look more carefully at how quantum probability works.

In order to define where we are with exploring science in our search for freedom, let's go to the year 1900. Max Planck (1858-1947) was working for the German Bureau of Standards on the problem of "black body" radiation. It's not actually about black bodies, but about how materials glow, or produce light, as they're heated up. The color of the glow of any hot object is not due to what the material is. It's only based on how hot it is. This was important for industry as light bulb technology was still being perfected. Planck's conclusion was that the energy involved is equal to some constant number multiplied by the frequency of the color we see.

This was expressed as $E=hf$, in which E is energy, f is frequency (how often the waves of energy occur, and therefore its color), and h is what we now call Planck's constant. We had found another precise scientific law. (Planck's constant is known today to be 6.626196×10^{-34} m^2kg/s, apparently an exact amount.) While by this time we knew many precise laws this one was different because h led to the discovery of a unique set of rules for the very small scale.

Five years after Planck's discovery, Albert Einstein (1879-1955) explained that Planck's constant (h) reveals something new about nature. It comes in minimum quantities, and h is the minimum quantity of energy. From this quantization we've been able to find that there are also minimum quantities of distance, time, mass, temperature, and more. All of these are known today to a very high degree of precision.

Uncertainty

What interests us for this investigation is that there's also a minimum quantity of certainty, as identified in 1927 by Werner Heisenberg (1901-1976). A popular version of his uncertainty principle comes to us in the form of

$$\Delta p \, \Delta x > h \, .$$

This means our uncertainty of momentum (Δp), which is mass times velocity, multiplied by our uncertainty of location (Δx), how big an area where we might find it, is at least Planck's constant (h). (Since its units are m^2kg/s we can use kilograms, meters, and seconds with this formula.). Even uncertainty is described by a precise formula.

The discovery of uncertainty seems to be our opening, our flexibility in the physical laws which gives us a way to get what we want because it makes all of our physical laws subject to this small amount of uncertainty. Heisenberg's formula seems to tell us exactly how big that window is. But let's see how it works and how it is that uncertainty, too, can be certain.

As we examine an object, we find there's a degree to which we can't be sure of its properties. This includes location. So let's calculate how certain we can be of the location of a baseball. If we measure its speed to be 50 mph (22.5 m/s) and its weight is 5 ounces (0.14 kg), then we know its momentum is 3.14 kgm/s. Therefore, its uncertainty of location is 2.1×10^{-34}m, which is about 13 Plancks. A Planck is the smallest measure of distance and it's unimaginably small.

This means no matter how precisely I calculate where it should land, I could be off by this tiny amount.

This doesn't give a baseball any significant freedom from the laws of physics. Physical laws don't appear imprecise enough to be considered flexible. But baseballs don't have a will, so they don't need freedom.

I believe I do have control over my arm, so let's see how much freedom my arm has. Any uncertainty in its location could allow me to put it where I like, regardless of where other natural laws say it should be.

If my arm weighs 5 pounds (2.27 kg) and I can measure its speed to be 50 mph (22.5 m/s), then its location is predictable to within 1.3×10^{-35}m. That's about 1 Planck of distance, or a billionth of a billionth of the width of an atom. That's not something I can work with. How can I throw a baseball if I can only deviate from the predictions of science by that amount?

My arm's movement is caused by muscle contractions. Those are executed by molecules of adenosine triphosphate, which store energy then use it to contract when they're told to. They get compressed similar to a spring, except using electrical repulsion instead of spring tension, and locked in place until released. They receive their orders from nerve cells which carry the signal from my brain. This is all according to natural laws. My arm doesn't need freedom to move as it's told. It just needs to be told. The signal for this comes from my brain. We do know it can arise either from an automated response such as a surprise or from a thought.

A thought involves millions of neurons firing together. A neuron firing requires a change in voltage of about 15 millivolts, or 15 thousandths of a volt of negative charge. This is very little force, but we need millions of neurons to acquire it, so this is what we need freedom to do. We need freedom to accumulate electrons, the carrier of negative electric charge for no physically predictable reason. Let's see what their uncertainty of location is.

An electron only weighs 9.1×10^{-31}kg. At slow speed it could have an uncertainty of position of 0.0001 meter (1/10 of a millimeter).

This may not seem like a large area, but electrons and their molecules are so small that there are likely a quadrillion electrons within reach. That means it would be reasonable, or within probability, to gather enough electrons to form a thought. We wouldn't need a physical explanation for those electrons being there. This looks like it could be the opportunity I need to form a thought and direct my arm to do what I want it to do.

Probability

The only true probability we find in science is this principle of uncertainty. And we can see by experimenting with its formula that it's only effective at the smallest scale. While every object does have an uncertainty of location or momentum which could be calculated, only subatomic particles are small enough to have any significant uncertainty to them.

Arthur Eddington (1882-1944) responded by saying that "religion first became possible for a reasonable scientific man about the year 1927", the year of Heisenberg's uncertainty principle. "We recognize a spiritual world alongside the physical world... the physicist now regards his own external world in a way which I can only describe as more mystical, though not less exact and practical, than that which prevailed some years ago."

Arthur Holly Compton believed quantum uncertainty implied "the possibility of mind acting on matter" and "the existence of a supreme being guiding the affairs of the universe, in which man probably represents the highest order of intelligence."

I could, at this point, suggest that this is where probability allows for a person's will to be implemented. These revelations are based on the limits which quantum probability gives to all physical predictions. It makes all of science a little bit uncertain. However, I don't believe this is the correct explanation because probability is a precise concept and has been proven so. While our world has this degree of uncertainty, it is not flexible because it still has a precise behavior. Probability allows for thoughts to form, electrons to gather, but not according to my will. It doesn't allow for intention, only chance. This is because while one particle's behavior is unpredictable, the behavior of a large quantity of them is predictable. And the more particles the more precisely their behavior can be predicted. Quantum calculations are statistical, as they analyze large quantities.

Quantum probability is actually the most proven law of nature we've found, since it's the basis of most modern technology. In the 1940s, Julian Schwinger, Richard Feynman, and Sin-Itiro Tomonaga formulated a method of defining precise probabilities for electronic applications. This is known as quantum electrodynamics. It's about the way energy interacts with electrons.

In 1947 scientists at Bell Labs used it to invent the transistor. Beyond use in radios, the transistor is a fundamental component of microchips. So microchips, and therefore almost all modern technology, work according to the predictions of quantum probability.

We prove this theory to be correct every time we use an electronic device. And when a device fails it's not because the theory has failed. It's because the device has a structural problem or interference from some other activity.

In addition to the success of this technology, the predictions of quantum electrodynamics have been proven correct with as much precision as we've been able to test for (which according to Toichiro Kinoshita's 30 years of work with it, is better than 1 part in a billion). And every indication is that probability is even more precise than that. We've found no reason to believe that it's anything less than absolutely precise.

In addition, all electrons are the same. Those in computers, those in trees, in rocks, and in your brain are all the same. Not only are they the same, but they're constantly trading places. The electrons in your brain have been in rocks and trees, and likely not too long ago, perhaps just minutes.

Probability tells us how certain we are, or how uncertain we could be of any scientific prediction. So let's take a closer look at how that is. We're actually going to be looking at complex situations since they're easier to visualize than particles. The reason is because any object which is big enough for us to see is too big for quantum probability to have any real effect on. And the behavior of some complexities replicate that of actual probability.

Flipping a coin is a good example of a complexity which results in the appearance of probability. If I flip a coin, the chance of it landing on heads or tails is 50% for each. We can chart this probability as:

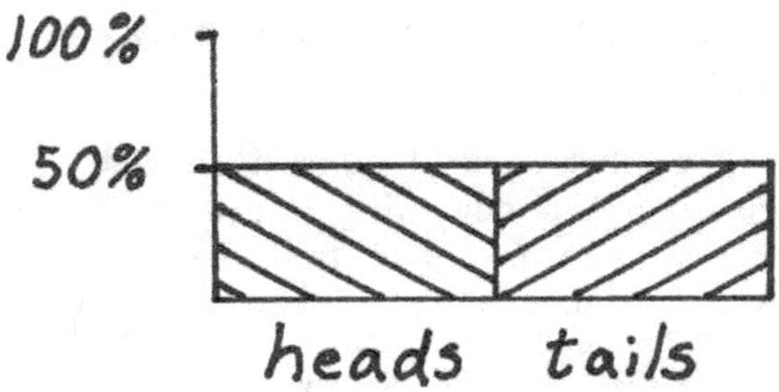

In a typical coin toss, I have no way to predict how the coin will land due to the extreme complexity of this seemingly simple act. One flip cannot be predicted. And yet its probability is precise. Its precision reveals itself in quantity. The more times I flip the coin, the more precisely I will get 50% of the total results to be heads and 50% tails. From 50% to 50.0% to 50.00% and so on with growing precision. Probability becomes perfect at infinite occurrences. This is the certainty which probability calculations rely on to make technology work.

While our real-world examples of probability aren't actual probabilities, they are good analogies for what happens at small scales. The only true probabilities in our world are quantum probabilities. Those only have a significant effect on the very small. Everything big enough for us to see is almost precisely deterministic, predictable by physical laws. This means that Laplace is right, for the most part, the world is predetermined but too complicated for us to predict.

Now imagine we reached into this coin flip experiment with our will and caused a few heads to come up. In that case, no matter how many more times we flipped the coin, heads would always be ahead by that amount. Our probability would be off. It wouldn't be perfect anymore.

Quantum probability works similar to a coin toss, but there are generally more than two possible outcomes. We can imagine these scenarios as a roll of a die. Supposing we had 6 possible outcomes, its chart could look just like that of a 6-sided die. For any other number of possibilities, it would be like a die with that number of sides.

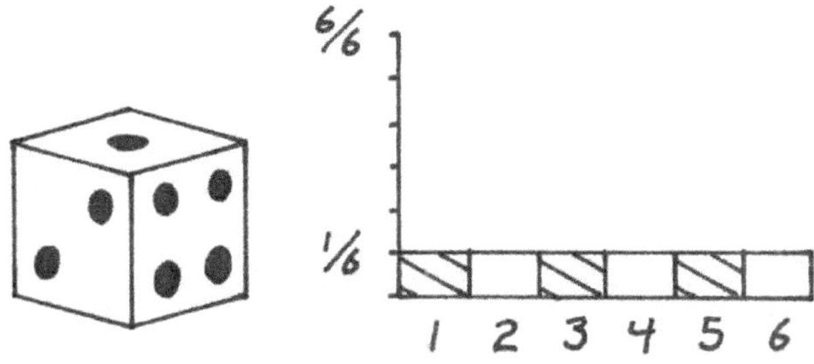

Each side of the die has the same 1 in 6 chance of coming up each time we roll it. We can't predict which will come up in 1 roll, but roll it enough times and we will witness the same precise distribution of results we see when flipping a coin.

We can apply this concept to every particle in our world. The state (the location, speed, etc.) of every one of the particles is not certain, but is a probability. And everything is made of particles, making the state of large objects based on these same probabilities. That means they have uncertainty, too, but their uncertainty is diminished greatly by the number of particles they're made of. An object of 100 particles has the certainty of a die rolled 100 times. We're certain of the properties of a grain of sand because it's made of a trillion trillion particles.

As I refer to particles, I include everything that is very small. The effect actually includes all things of any size, whether it's an electron, photon, human, or planet, but probability reveals itself more the smaller an object is, as we saw with the electron compared to my arm. It's like one roll of a die compared to many. So we can generally ignore probability as it applies to large objects.

Anything large enough for us to see has an extremely reliable behavior. But mathematically, probability is still there.

Disturbed by this discovery, Einstein is known to have said of God that "I am at all events convinced that HE does not play dice."

Particles and Waves

In the early stages of development, quantum discoveries seemed quite radical. Photons, particles of light, were found to have an uncertainty of location even though we all know light goes straight. If we tried to narrow light's path to be straight, it may choose some other route. Furthermore, if an experimenter tried to determine which path a particle took, when given two equally likely paths, then the object took only one. If he tested only one of the two paths, it took the one he tested for. Then if he didn't test for either path, the particle behaved as if it took both. It interfered with itself when both paths recombined at a target.

A single object, even a fundamental indivisible particle, given two paths to get to the same destination seems to take both of them as if it had split into two objects, then the two objects interfere with each other upon recombining as if they are waves.

Aside from uncertainty of path and location, another dramatic realization was that objects can be both a particle and a wave, and which one they are, seems to be determined by the observer. If a scientist did an experiment to find out if something is a particle, he always found that it was. And if he tried to determine if it is a wave, he also found that it was.

(J.J. Thomson received the Nobel Prize for proving that electrons are particles. His son George received the Nobel Prize for proving that electrons are waves.)

This led Niels Bohr and Werner Heisenberg to create what's known as the Copenhagen interpretation of quantum theory in 1927. It simply was that the state of an object is determined by the observer's consciousness. More properly, the observer's consciousness causes the wave function to collapse. The wavefunction is the probability of what the state of the object could be, and this may be imagined as a wave stretching across all of the options.

Possibilities When Not Observed   Definite Properties When Observed

This means all objects have all possible states (they are everywhere and are doing everything) until they're observed. Then they pick just one and all the other possibilities disappear. In the case of split paths, they're only considered to be observed when they hit a target since we can't see them travel. This doesn't mean an observer can always choose where an object will be, only that the object behaves "normally" when looked at, and has an undetermined behavior when no one is looking.

This led to a lot of speculation about potential metaphysical (non-physical) involvement and study of consciousness which continues even today. But among physicists, it mostly became set aside by a drive for technological advancement with the coming of World War II.

Then, in 1957, Hugh Everett proposed his many-worlds interpretation as an entirely materialistic approach, which is still popular today. This is a model in which every possibility exists in however many worlds there needs to be to accommodate them all. Everything which could happen does happen in some world.

In this interpretation the probabilities don't go away; they split into additional universes every time a choice is made. Objects maintain all of their probabilities until an interaction occurs, such as with a detector in an experiment. Interacting with a detector causes a particle to select that one state (such as a position or path) and every other possibility which existed prior to the interaction goes on to exist in a new universe of its own.

This means every time you encounter a choice, such as turning left or right, there is another universe created with another you, the you who made the other choice.

A popular current version of the many-worlds interpretation of quantum probability is the multiverse.

I offer another solution which is integrated into my theory of movement. It is that all the probabilities do exist simultaneously, but only one is evident in our world at any one moment. This may sound similar to many-worlds, except that the other probabilities are not the same as outcomes. Only one outcome, or reality, exists. The probabilities may be viewed as potentials in this model. They don't exist physically. And regarding whether an object is a particle or a wave; it's both. Everything is a particle when it's in space, but it's a wave as it relocates, which is frequently because all things have some degree of motion. We'll look at this in greater detail when we get to movement theory.

While the quantum paradox continued, the radical possibility of just anything occurring was turned into a probability curve. And sometime later Joseph Ford of the Georgia Institute of Technology was able to respond to Einstein, saying "God does play dice with the universe, but they're loaded dice." This doesn't bring certainty to any single prediction, it just changes our chart and gives us more focused bulk predictions. Any particle may still be found in any of the allowed states (any location in the universe where there is not already a competing particle, any speed less than light, etc.), but we have a probability curve for those states rather than the even probability of a straight line. Our probability for particles looks just like that of a loaded die.

We still don't know what state we'll find any one particle to be in, just as we don't know how a loaded die will land. But, with enough rolls, or by looking at enough particles, we'll find a perfectly formed probability.

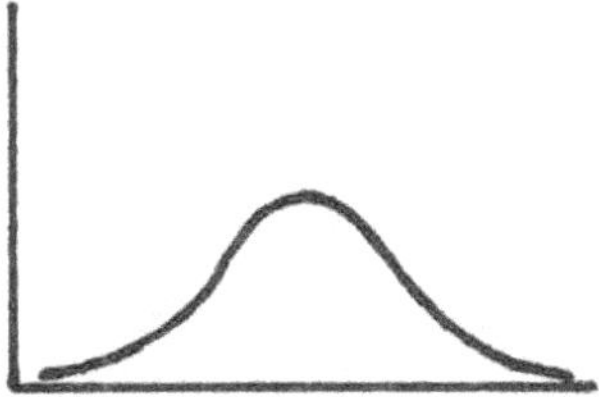

This means that we do have a reasonable idea of where a large object is or how fast it's moving, even though there is some small chance it could be otherwise. The low portions of the curve extend indefinitely and never become zero. This is how extremely unlikely events do and must occur, such as radioactive decay and mistakes in DNA duplication. (An atom of Uranium only has a 0.000000001% chance of emitting radiation this year. A nucleic acid pair in DNA only has a 0.0000008% chance of error when DNA replicates.)

And, as with our coin flips, any interference would throw off this perfection.

While one object measured once has a degree of unpredictability, as I calculated with the baseball, my arm, and electrons, many objects or many occurrences form precise probability curves which couldn't be perfect curves if there was any manipulation of them.

Radioactive decay of one atom is unpredictable, but decay within a very large quantity is very predictable. This is how we know half-lives of elements. Flaws in DNA duplication are similarly unpredictable in one instance since duplication should be perfect. But we've learned that since humans are made of so many cells (up to 100 trillion), and DNA contains about 30,000 genes for a total of about 3 billion base pairs of nucleic acids, each of us bring about 100 mutations into the world when we're conceived.

I think the precision of probability rules out the possibility of consciousness being a factor in the behavior of objects.

In the certainty of probability, we've lost freedom in the physical world, but probability gives us something in exchange.

There's a reason why quantum behaviors must be probabilistic, and it has to do with another principle of nature which provides the world with a certain destiny.

Chapter 7 - Entropy

Entropy is a measure of the disorder or randomness in a group of objects. In 1865, Rudolf Clausius introduced the idea of entropy as the disordering of a system. A system can be any group of objects which are physically related in some way, such as the molecules in a ball. Disorder is a lack of meaning to an arrangement. With enough disorder, a ball is no longer a ball.

The second law of thermodynamics tells us that for any system, entropy always increases. This is often referred to as one of the arrows of time, how we know which direction is forward time.

Henri Poincaré (1854-1912) found that the probability of entropy decreasing for a system is 1 in 10^n, in which n is the number of particles in the system. So, a system of 10 particles would have a 1 in 10^{10} or 1 in 10,000,000,000 chance of becoming more ordered. (The number of particles is the number of zeros.) This method of estimation tells us that the more particles an object is made of, or the more particles which are involved in a situation, the more likely it will deteriorate at every opportunity. (An opportunity is any event, any tiny movement, even vibration from heat or quantum fluctuations.)

I pointed out previously that the greater the number of particles, the more certain we are of a system's properties. Here we see that, along with that certainty of the whole object, how sure we are of what something is and its behavior, we also get deterioration. The more particles something is made of, the more sure we are of what that something is and is doing, and the more sure we are that it will decay a little with every moment.

Our universe, with its 10^{87} particles, has very certain properties overall (making it what it is), but has no realistic chance of becoming more ordered, as a whole. It must decay continually.

Our bodies, with their 10^{28} atoms, also must decay. The chance of our bodies not decaying in any particular moment of time is 1 in $10^{10^{28}}$ or $10^{10,000,000,000,000,000,000,000,000,000}$.

Complexity vs Quantum Probability

There are two basic principles which create disorder, complexity and quantum probability.

- Complexity can look just like probability, such as in the case of a coin flip, but actually just involves a lot of variables.

- Quantum probability is true probability. While a coin flip is too complex for us to predict, the quantum state of a single object cannot be predicted by any intelligence.

Disorder is an arrangement of objects which has no current purpose. It may have had a meaningful arrangement in the past but no longer does. If its past logic is the result of complexity then the current arrangement could have been predicted, and so could the placement of each object or particle.

If I dump a bucket of sand, I know what shape the pile will make. At a large scale it is ordered. But predicting the location of each grain is too difficult. At a small scale it is disordered. The arrangement is disordered if it has no specific meaningful placement for each object.

If the current disorder is the result of quantum probability then it too, has a past logic, and that logic is probability. Its current arrangement could be predicted as a whole, but the placement of each object cannot be predicted.

If I project a beam of light (photons) at a wall I know what the arrangement will look like, but it's not possible to predict the destination of each photon, each particle of light. I can only determine the probability of them landing where I aim.

The placement of each object may have resulted from a purposeful logic and may be predictable, as with a complex system. Or the placement of each object could be the result of quantum probability and be unpredictable. Both are disordered if their current arrangement lacks meaning.

The quantum probabilities, as we discussed, provide a true unpredictability which Laplace cannot overcome with any amount of intellect or knowledge. His proposed superior intellect could predict how a rolled die will land but not how a particle will.

Only one form of disorder blurs our view of the future. Although both lead to a general disordering of our world, only one assures its deterioration. It's not the complexity of a rolled die, it's the unknowable destination of a tiny particle.

Complex placement does not cause radioactive decay or mutations of DNA. Only quantum probability can do that because it can cause particles to go around other physical laws, such as an electric bond or nuclear bond within an atom. In this way only quantum probability can cause true deterioration of our world. It removes prediction from the realm of possibility. So, all true decay occurs at the subatomic (particle) level of our world.

Because of probability, or because of the expanse of the probability curve, we have an assured and steady disordering of the universe defined by the shape of that curve as well as an unpredictability to our future which increases the further we try to look ahead.

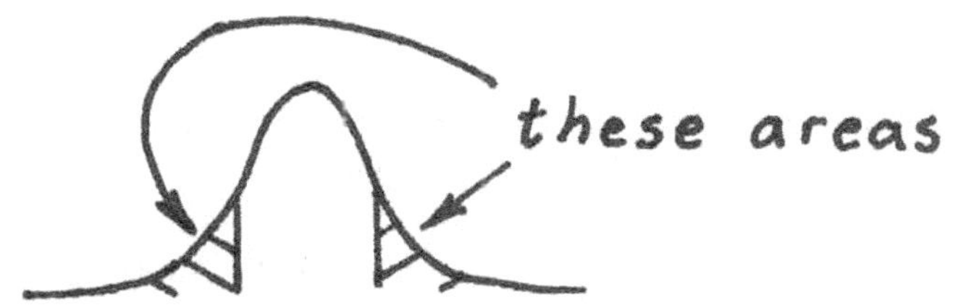

In other words, our world is moving toward decay because of quantum probability, and our view of the future is nearsighted.

This raises the question of God's ability to know the future. In many religions God is all knowing. It may seem that this could limit our freedom. I think, here, we need to reconsider the conflict of Laplace's statement (that our mechanical world is predictable) with free will. If God knows the future, then it doesn't seem to make sense that we should have freedom to act as we choose. And, without freedom we don't have the responsibility to make good choices, to behave according to our religious obligations.

I believe there's a way to understand how God can know the future while it remains uncertain, and we have freedom. Since God's not physical he's not in time. We are both physical and non-physical and these aspects of us seem to be bound to each other. So as humans we experience the world in time. The future is blurry to us because of our inability to calculate complex outcomes and also because it actually is uncertain to someone in time.

The omniscience of God (his all-knowingness) is a difficult concept to understand since we like to think that he experiences time as we do. Because physics is bound to time and God isn't, his knowledge of the future doesn't affect it. In this way it's possible for him to know the future while it remains uncertain for us, and we remain fully in control of our choices. His knowledge outside of time, outside of physics, does not impede our freedom within the physical world.

The result is that we are actively involved in the world. But it doesn't mean he isn't. While His omniscience doesn't allow him to be bound to the physical world, he is known by some religions to use actors to carry out his will within time and space.

If all physical events were to occur precisely every time, then Laplace would have been correct, and the future would be precisely determined. As it is, with precise probabilities, it can't be known (when viewed from within time), nor can it be affected. Probability is a very special tool for giving us a perfectly functioning and logical set of laws, yet not allowing anyone living in time to see the future.

It also assures us of decay. The high part of the bell curve of probability gives us the degree of certainty which we have and keeps all laws of nature working. But the low parts which extend indefinitely are what disorders and breaks down the world. Those slim extensions are the reason why decay occurs:

- Carbon 14 has a 50% chance of losing part of its nucleus in 5570 years.

- Uranium 238 has a 50% chance of losing part of its nucleus in 4.5 billion years.

- A neutron which is outside of a nucleus has a 50% chance of decaying in 13 minutes.

- Protons, which are the key part of every atomic nucleus are expected to decay in 10^{34} years.

Everything in our world may someday break down into its fundamental component of energy (light).

Many people believe entropy gives time its direction too. It is pointed out that we remember the past but not the future. Because of uncertainty causing entropy, we can only see the future with limited clarity. As the whole universe is, by law, destined to obey probability which results in disorder, the forward direction of time is distinct from the backward direction. While history is perfectly clear, the future is fuzzier the further forward we try to see.

Entropy Wins

As I walk along the beach and see a mound of sand, I know that tomorrow the mound will be lower. Piling sand is an ordering process because it creates something. The deterioration of the mound is disordering. Acquiring meaningless placement is disordering.

Probability and complex placement result in disorder because of the many ways things can be arranged. More ways are without meaning than with meaning. When change occurs, whether by a complex event or by probability, a system may become more ordered or disordered. Because there are more possibilities of a disordered arrangement, the likelihood of disorder exceeds that of order. So whenever an event occurs it usually results in disorder unless there is a force involved in the event causing order.

Order can and does happen by complex events based on common physical laws and by quantum probability, too, but not as often as disorder. Because events keep happening, order is temporary. And with quantum probability, it is its own cause of events.

Quantum fluctuations have no known cause. They just continue to happen. The imprecision it gives the destination of a particle continues to cause the particle to relocate even when no cause is present besides probability. If we don't touch an object, it still deteriorates with each passing moment because probability is always relocating particles.

Arthur Eddington wrote, "The law that entropy always increases, holds, I think, the supreme position among the laws of nature."

The ordering of our world, as well as the unstoppable disordering of it, both occur at the particle, or subatomic level.

Objects join together by the forces of gravity, electrical attraction, and nuclear bonding. These same forces pull objects together to form large objects like planets, stars, and galaxies, and also small ones such as crystalline structures, molecules, and various atomic elements.

While complex interactions use these forces to move objects around, quantum probability breaks these bonds. Quantum probability takes apart advanced objects, such as elements, which are held together by the strong nuclear force, the strongest force in our world. It allows particles to escape from inescapable environments, including black holes. No force of nature can hold together a group of particles well enough to prevent probability from dismantling it.

Quantum probability is able to do this because it is not a force. It doesn't pull a particle out of its imprisonment. It causes it to disappear and instantly re-appear somewhere else. No strength is required.

It is, as Arthur Eddington pointed out, the ultimate principle of our world. And it is the world's undoing.

Many people look to the ordering tendencies of nature for answers to questions like how life formed, how it advanced, and why we seem to be independently thinking and wanting beings. But the numbers are against them. The probabilities point to devolution and decay, not evolution and advancement. Both complexity and quantum probability work together on this. But we'll see how quantum probability offers a particular opportunity for a benefit to living things while assuring ultimate deterioration of all things regardless of what physical laws may temporarily create order or resist disorder.

While the world at our large scale looks orderly and should be predictable, everything which happens is a result of activities at the subatomic level. All the big things we do are based on the behavior of particles. So if we want to do anything in this world we must do it at the subatomic level.

Chapter 8 - Evolution

Entropy gives our world a steady march toward devolution.

We often hear of evolution, how our world got to be the way it is. It certainly has turned into something more sophisticated than it was. But evolution is not a principle of our world. It's a series of events which have happened but are not all due to physical laws.

Chemical compounds necessary for life are a good example of physical laws working toward evolution, but the assembly of those into lifeforms is unexplainable by nature. Electrical attraction between atoms assembles them into molecules, but assembling those into cells requires a very specific and unlikely arrangement of molecules. Physical laws, all the laws of nature we've been able to discover, not only do not support much of the evolution that's occurred, but instead explain how it gets undone.

Humanity has worked to discover principles it can use to create technology. This may be looked at as evolution of our world, advancement of civilization, but not of the human body or anything natural. It also creates more disorder than order, as every process does.

Every process which orders our world also creates a greater amount of disorder. Every product we make results in more disorder in the form of waste material and energy loss, such as heat, than the total order of the product we make. The resources needed to build a house far exceeds the resources needed to knock it down. This is not because we're wasteful. It's the way of nature. It's those ends of the probability curve showing themselves whether it's complexity or quantum probability. As we order our world, we create more disorder than order.

A star manufactures chemical elements by pushing protons and neutrons together into bigger atomic nuclei (the center part of atoms). In the process, it produces a far greater disorder in the form of light, heat, and other radiation and waste particles. The heat is disorder because it's emitted at random frequencies and in random directions.

An animal consumes and disorders more material than its body is made of. Part of the disordering occurs during the process of growing. So, its disordering is both complexity and quantum probability.

Pressures and temperatures in our atmosphere overcome disorder and create clouds, rain, tornadoes, and other patterns in the air.

These ordering phenomena are the probable result of our laws operating over time. We can't predict any individual particles to fuse or individual raindrops to form due to quantum uncertainty and complexity. We would say they are the result of chance, similar to how rolling a loaded die is to those of us who aren't able to calculate its result.

Many people believe the existence of you and me is also a result of chance, that humanity and all life is a consequence of the universe's progression toward ultimate disorder during which physical laws create occasional patterns. But what is the chance of that?

If I gave you a piece of coal, would you squeeze it with the hope of forming a diamond? It really could form a diamond, but you wouldn't squeeze it because the probability of achieving that manually is so extremely low that it wouldn't be worth doing. You'd be applying a little pressure and waiting for random movements to create order.

Even lower is the probability of forming you and me, or any living thing on this planet, in the galaxy, or in the universe. In fact, a universe of universes (what's referred to today as a multiverse), has no reasonable probability of forming any living thing.

The Multiverse

But let's assume that it could and did. A huge number of universes, by chance, formed a living organism. That creature would have to, by chance, evolve against nature's devolutionary tendency in order to get us to where we are today. More improbability. And we haven't even explored the improbability of having the precise laws we have which allow for life to exist.

In a random creation of universes, the chance of selecting laws which allow for life is improbable to a massive degree. Of course, the concept of creation itself, the formation or existence of any universe is also unexplainable by any physical methods. Physics can't explain how to get something from nothing.

Practical Impossibility

It is unlikely that the world as we know it could exist. Setting aside issues such as life, consciousness and will, and the existence of matter, and considering only the assembly of physical bodies when we look at probability, it's easy to say that anything could happen and obviously it did, because here we are. But that doesn't acknowledge the concept of practical impossibility. For example, there could be an elephant in my living room. Sure, it's possible, but not really. We know what makes sense. We know what can and can't happen in a reasonable way. This was Einstein's concern about uncertainty, that it does allow for anything to happen. However, the ability for anything to happen is tempered by degrees of possibility.

You may have a one in a million (10^6) chance of winning the lottery. It's not very likely but it could happen. But, putting some material in a jar and shaking it to form a human brain cannot happen. According to Ludwig Boltzmann (1844-1906), it would take $10^{10^{68}}$ years just to have a reasonable chance. That's 10 to the 10^{68}, an exponent of an exponent. If would take 250 billion pages to write it in normal form. Shaking a jar of material to form a human brain is a practical impossibility.

John D. Barrow and Frank J. Tipler found that the odds of randomly assembling a human genome is 1 in $10^{12,000,000}$. Even this number would take 4300 pages to write in normal form. If you started out with some life form which already had DNA, you would need that many mutations to randomly create a human DNA molecule.

But only 10^{40} organisms have ever lived on earth, so only 10^{40} attempts could have occurred in developing a human. And we still need to somehow create DNA to have those organisms.

Paul Davies estimates the chance of DNA itself assembling to be 1 in 10^{4000}.

Additionally, only 10^{60} moments of time (Plancks, the smallest measure of time) have passed since earth formed, so time is tight too. But let's apply that to every particle in the universe just to see what that looks like. It's estimated that there are 10^{87} fundamental, indivisible particles. If every one of them had an event at every moment since the big bang there would have been 10^{147} events. (These aren't large events like a roll of a die or a cell reproducing. This is each one of the smallest bits of our universe moving the smallest amount.) The maximum number of events which could have happened so far in the whole life of the universe is 10^{147}. You could suggest multiple universes to answer these questions but the number of universes required would be beyond unimaginable. And what would be the explanation for that?

Somehow life was formed, and not by a natural tendency to come about. If there was some undiscovered physical law creating life it would continue to occur. We know life only formed once because we all share similar DNA characteristics. Every living thing.

In addition to common DNA structure, every living thing is made of left spin particles. All particles spin, and we call them either left or right-handed. Their direction is randomly selected as they're created, but as they interact they need to be compatible. For chemical reactions, which is how all living organisms derive energy, they must have the same spin. In our world they're all left-handed. If we ate something made of right-handed spin particles, we would not be able to digest it. This trick is used to provide us with artificial sweetener. It's real sugar but has the wrong spin. In this way we can't chemically interact with it, which means we also can't get energy from it, nor can we gain weight by eating it.

When life formed it had a 50% chance of having a left-handed spin. If it had formed more than once, each time it would have a 50% chance. This tells us that life only formed once. Had it formed multiple times, chance would have caused half of all animals and vegetables to be undigestible to us.

Earth formed about 4.5 billion years ago. Life originated once about 3.8 billion years ago from non-living matter and has not formed from non-living matter since then. It seems that something intervened somehow.

I believe the brilliance of Charles Darwin (1809-1882) may lie more in his identification of an intervening will, than in his explanation of the speciation of creatures, as he wrote about in *Origin of Species by the Means of Natural Selection*. He is popularly known for discovering that the refinement of a kind of lifeform occurs by random mutations being selected by survival or non-survival, based on whether the mutation is beneficial or not. A creature born with a mutation which benefits it will survive to reproduce and create more with the same mutation. One with a disadvantage won't live to reproduce and pass on its genes. This is commonly accepted as the method by which nature creates differences within a type of animal. They may have adaptations to their environment, such as colors, sizes, and other variations which make each species of a kind of animal unique.

The driver of this speciation, I believe, is an equally great discovery. Darwin found that every creature, everywhere, which he came across in his 5-year journey around the world reproduced excessively. Each one created far more offspring than its environment could support. Had all the offspring of all living things found enough food, the earth would have been overpopulated very quickly. Every habitat would have been overrun right away.

It doesn't seem that any living thing would do that. Reproducing comes at great cost for every type of creature. Every mother gives a tremendous amount of her own nutrients and health to her offspring. She also gives greatly of her safety during pregnancy and child raising.

It seems that self-preservation should be a creature's highest priority. But every creature produces offspring that could die young of starvation, if for no other reason, and at great cost to itself. In fact, it's surprising that creatures want anything at all, if they obey the laws of science. Every living thing has a desire to make more of itself in excessive amounts, and that seems to supersede any other desire.

One could suggest that the desire to reproduce is itself a feature which got selected. However, reproducing excessively works against the survival of both parent and child and its only benefit is faster speciation, which is still a very slow process involving multiple generations. So, it's unlikely to get selected. Faster speciation provides no benefit to the parent. Yet every creature is very forward looking. And what is the physical mechanism of wanting? I suggest wanting is a non-physical cause.

Darwin suggested that the same small steps which provide for speciation may eventually lead to a new kind of creature. But he also knew there was no evidence of the in-between, or transitional, creatures so he admittedly never found the origin of the kinds we have today. However, he did point out that they all appear to have a common ancestry due to similar features and must somehow have been born to each other. He described a natural device for adaptation in our devolving world, and also discovered the presence of will and found it a necessary ingredient even in his own theory of speciation.

Since his time, scientists have explored the possibilities of mutations being behind evolution, getting us from blue-green algae to fish to rabbits to apes to humans, and of course all the steps in between. And they haven't found that it's been able to work.

Mutations have been caused by radiation or chemicals in laboratory animals for decades with no sign of advancement. None of the mutations have benefitted the offspring. Occasionally they have been harmless. Mostly they have been harmful, shortening the creature's life or reducing its ability to survive. Never has it been an advantage. And never has there been an indication that one kind of animal could turn into another kind. An ant can't turn into a fly. A fly can't turn into a wasp. A mouse can't turn into a rabbit. And the type of animal which it is, seems not to be based on its DNA, but rather on the cell structure it gets from its parents. That cell structure appears to be immune to mutation. This immunity is likely due to it being of far larger scale than DNA, which is a molecule. Being of larger scale doesn't make it fully immune to mutation but far more resistant to the effects of quantum probability which are usually tiny.

In the 1960's, Murray Eden found that random mutations to genetic code could only deteriorate current functions.

Christiane Nüsslein-Volhard and Eric F. Wieschaus won the 1995 Nobel Prize for Medicine or Physiology by studying the mutation of fruit flies, discovering every genetic aspect of their body plans. Every one of their mutants died before adulthood and none of them formed a new type.

Meanwhile, we've also learned that all things we've found to have ever lived on earth share common DNA. We all have a common ancestor.

Human DNA is

- 98.5% the same as chimpanzees,

- over 90% the same as mice, and

- over 60% the same as fruit flies.

It seems that each new kind, since they're all related genetically, must have been born to another kind.

Archaeology

Archaeologists build evolutionary diagrams to form a family tree of all life on earth. While we, admittedly, have only a small percentage of creatures identified (according to Richard Leakey and Roger Lewin, 30 billion species have lived on earth, but only 250,000 have been found as fossils), there are certain types of creatures conspicuously missing. Those are the links between kinds.

Darwin thought that the major kinds (phyla) of creatures could have come into existence gradually. the same way speciation occurs by random mutation. However, fossil records now show that 20 of the 27 phyla came into existence within just one 5 million year period in the 3.8 billion years since bacterial life first formed on earth. That period was about 530 million years ago and is known as the Cambrian explosion. Five million years sounds like a long time but it represents only one tenth of one percent of the time life has existed on earth.

Looking at evidence, it appears that our world evolved. But we seem to be told by religion that it didn't evolve naturally, and natural laws tell us it shouldn't. There's a direct conflict here.

We're told that creation culminated with the birth of humanity. Studying historical evidence, archaeologists have assembled the steps of evolution, the creatures in order of their occurrence. We're told creation occurred in steps. Likewise, evidence shows us we "evolved" in steps.

And after the appearance of man there's no mention of further creation, nor is there evidence of further evolution, according to scientists. What there is, though, is speciation. And according to Nathaniel T. Jeanson, mitrochondrial DNA tells us that the species we find in our world are no more than several thousand years old. (Interestingly, humans did not speciate in the 45,000 years since their beginning.)

We can see from all of this that nothing about natural laws provides a reasonable possibility of life becoming more advanced, even if we assume it could come into existence on its own. In fact, free will intervening from the outside is necessary to explain how creatures became specialized, and that's not even advancement. The specialization, or speciation, of life-forms is a lateral move which the will to reproduce excessively has achieved.

We see that quantum probability makes all predictions of physics a little imprecise. This imprecision occurs constantly so it does allow for anything to happen. But it only works by chance, not intention. If we interfere with a coin flip then the probability is disrupted. Likewise, if we interfere with the random probabilities which are involved in the world, even in our own bodies, then probability would be off.

While anything can happen, it can only happen by chance. That isn't able to accomplish evolution or the execution of our own will. My arm can make movements predicted by physics, such as an automated response from my brain to being startled. Probability also allows it to make random movements against predictions. But it cannot make intentional movements. The world is predetermined, but its precision seems blurred by quantum probability. That's not freedom.

We're still left with how to create the advancement in life as fossil records show us, and how to implement our own will. We need new information to be added for that. So let's look more closely to see what's needed. Then we'll return to our blurry world to see how we can have real freedom.

Chapter 9 - New Information

With all the probabilities being precise, how did we get here, and how do we continue to overcome the probabilities to implement our will? New information must have been introduced for creation to occur, and we must be introducing more new information to advance and live our lives. We'll soon see how to put new information into our precise world, but first we'll look at what it is which needs to get through our physical obstacles.

New information is creativity. It's an idea, a design, or process, which wasn't previously known. More specifically, it's any purposeful act which is not caused by natural laws, not even the randomness of quantum probability.

New information doesn't have to be a new design for a lifeform or even a feature. It doesn't have to be a sculpture or hairstyle which has never been seen before. However, it must be something which cannot be predicted or expected, but is purposeful. New information is the fundamental component of thought. It is what makes thought different from calculation. Thought leads to a choice. Calculating leads to a result.

While a computer can ask a question and determine its answer, it cannot create the question. Questions must be created and given to a computer or calculating device.

A machine could be made that is able to determine the weather, fly a plane, or even build a plane. It can encounter problems and solve them. But the process of navigating obstacles requires asking questions. Determining solutions requires creating options for what the solutions may be. A machine cannot make up its own questions to be asked or its own answers or solutions to implement.

If a robot runs into a wall, it can determine that there's a wall there if it asks itself if there's a wall there, and if it knows what a wall is. But the robot cannot form the question it asks or the answer it selects. A programmer must do that. It also cannot decide on its own to ask a question. Someone must give the robot a trigger to cause the asking of the question.

The laws of nature are logical. We would say they make sense once we understand the reasoning behind them. Even probability, the chance fluctuations at the subatomic level which cannot be known in advance, are logical. Although they can't be predicted individually, they can be predicted when considering a large number of occurrences.

That doesn't mean that new information is illogical. It can be new logic. Creativity is something which is new.

Biologists have not found that mutation can bring new information into the world. They haven't found any sign that a natural process can create new kinds of creatures or new features such as jaws, wings, scales, or fur. Lee Spetner of Johns Hopkins University has said, "... of all the mutations studied since genetics became a science, not a single one has been found that adds a little information." The reason is because mutation causes mistakes and the chance of a number of beneficial mistakes all occurring at once is a practical impossibility. New information is more complex than the mechanism of natural selection used in speciation.

The laws of nature were written before we got here. The logic of this world already existed. We've found no way in which logic may write new logic. The structure of logic is that all of its layers, everything which may come from it, is inherent in the fundamental logic. All that arises was pre-existing.

We can see this in the sciences. The logic of chemistry is how biology works, but the logic of atomic physics is how chemistry works. The laws of chemistry are built into physics. Everything about chemistry is the way it is because of the way physics is. There is nothing about it which is not entirely dependent on physics.

The whole logical structure of the world works this way. So whatever the most fundamental laws of nature are, all of the laws of nature are ultimately built into them.

In the early universe, as atoms formed, no new laws needed to be put into place to tell them how to behave. When gases gathered and condensed by gravity into stars and planets, the principles which dictated that process and those new objects were built into the original laws of the universe. As the first star formed, there was no new design. The stars, the planets, rocks, and air, were designed into nature from the beginning.

But not everything was. It's those unexplained moments of creation that we're looking to explain. The big ones like the creation of man, and the small ones like movie scripts and why we choose our second favorite meal at a restaurant rather than our most favorite. Though the events are dramatically different, they each require something new.

Invention

The design of the wheel can be said to be inherent in the laws of nature. We have gravity and we have laws of mechanics, such as friction and leverage. And when we apply those mechanical principles in a certain way, we get a device to help us work against friction from gravity.

The wheel reduces the friction of moving a heavy object across the ground by using leverage to reduce the travel distance of the parts involved in friction. When we roll a cart a significant distance, the wheels allow us greatly reduced friction while traveling across the ground because they act as a lever. The wheel reduces several feet of travel to just a few inches of axel rotating.

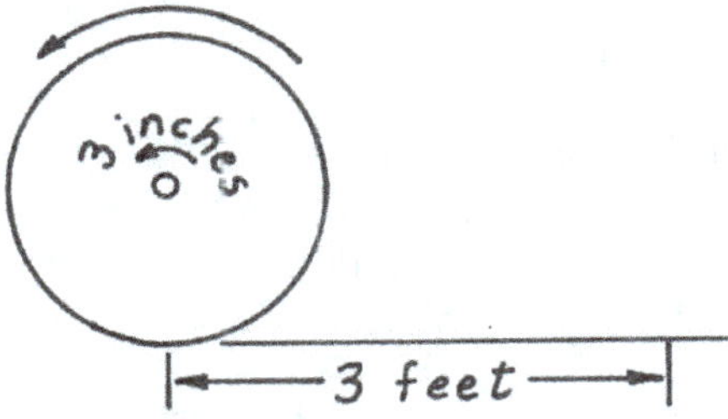

In this example we see a travel of 3 feet costing only 3 inches of friction rather than the 3 feet of friction we would have if we had dragged the load.

Sometimes inventions are thought of as discoveries. An inventor may be said to have discovered a way of reducing friction of travel. And certainly, a worthwhile invention such as that is likely to be made due to the way physical principles fit together and the number of people attempting to discover a way to accomplish work more easily.

Discovery

This is similar to scientific discovery. It is said that Einstein's special theory of relativity was bound to be discovered by someone because of Maxwell's discovery of how light travels, and Lorentz's discovery of distance adjusting itself to accommodate light, along with Reimann's geometry which allows distance to adjust. Someone was likely to put it together into a theory of relativity, a fundamental theory of physics.

And along with this train of thought comes the popular belief that there is nothing new under the sun, just new ways of applying old ideas. But where did the old ideas come from so inventors could apply them?

While invention implies creativity and discovery may not, both are creative acts. Calling an invention a discovery of how physical principles can be combined does not subtract from the accomplishment. New information was still brought into the world.

Since our world is made of different components, new information can take the form of a different and useful, or potentially useful, assembly of those components.

This can be demonstrated using a set of Lego blocks. Designers provide us with plans for useful objects we can assemble the blocks into. Or we may be a designer ourselves and create our own unique objects out of the blocks.

In our world we have specially designed objects which are able to produce other specially designed objects. They're called cells. But cells are like factories of robots which assemble molecules according to instructions. Due to quantum probability at that scale, they can make mistakes resulting in a less effective product, and they could possibly make a mistake which specializes the product slightly. But they cannot create a whole new useful object without creative thought.

The designer of our world has that creative thought. Humans have it, too. While the creator had the whole universe to work with, we have our bodies and our local resources to use. This is how we're able to advance technologically and become what we are as modern humans. We've created social structures and physical structures, and lots of useful objects to make our lives easier and more enjoyable. A non-creative being couldn't do that.

When a human assembles Lego blocks, he may follow instructions which another human has provided. Or he may create his own designs. He doesn't come up with useful Lego assemblies by years of random trial and error. He does it within minutes. He has an idea and builds it with very little error. This new design is creativity, and it is entirely unexplained by any physical principles. It allows us to make the practical impossibility of random trial and error into something very possible. It allows us to overcome probability.

We may also look at scientific discovery in a similar way. Scientist's look at a phenomenon and speculate about how it may have occurred. The way it occurred is a pre-existing logic. It's not new. But since we don't know what it is we create a new logic and try to determine if that new logic of ours, new information we've introduced, explains or predicts the phenomenon. If it does, then we may have discovered the pre-existing logic which the creator wrote. Looking back on discovery, it may seem easy to deny that it's creative, but the true process of scientific discovery cannot avoid scientists creating new logic to test. Scientific discovery cannot be done without creativity, which for them is the introduction of new information in the form of a hypothesis.

Where It Comes From

My theory of movement identifies a need for a non-physical dimension to allow for movement. It also requires the objects in the non-physical realm to interact with space and time. It's easy to see how a constant connection of our world to a non-physical existence tells us that our will is able to exist outside of the physical world. In fact, it needs to because new information must originate outside of space and time. Due to the strictness and constancy of physical laws, creativity cannot originate in the material world.

But simply acknowledging the non-physical aspect of this will doesn't explain how it's allowed to cause things to happen. It only acknowledges that our will can exist and that our wants may not be subject to physical law. But our body's actions still are.

We are more than just our bodies. Likewise, there is also more than just space and time. Our universe, too, is more than it appears to be.

At this point we have physical dimensions with strict laws and also a non-physical aspect of our world and us. The non-physical, though, is subject to the physical laws as well, any time it tries to affect our world. So the <u>us,</u> which is not physical, is free to want, but must also obey laws when it acts.

There is a way to put new information into this strict world. We've been given an entirely logical world to live in, and the ability to do what we want as well. Seeing how this works will take time.

Chapter 10 - Time

I believe a common mistake in trying to define time is to refer to the clock itself as the determiner of time. This leads to the idea of the events of the clock's workings being the underlying structure of time, that time emerges from the ticking.

Many scientists today believe that time is such an emergent phenomenon, that it's not fundamental. Some might even say it's not real or it doesn't exist. A popular explanation is that it results from a series of events that, like the events within a clock, cause time to appear to advance. These events are the fundamental components of nature, and what moves us forward giving us the experience of time advancing is those events.

This would mean that the rate of time is determined by how fast those events occur. If different events occur at different rates, then time is irregular. And if there was no event, then time wouldn't pass. It would seem that the more steps a person took, the more time he would experience. But this is the opposite of what relativity tells us.

The relativistic effects which Einstein explained to us and which I maintain in my theory of absolute relativity, are that time is reduced with speed (and space is also). This means that less time occurs with more events. My explanation of how this is involves showing that events, changes, occur outside of time. In this way experience of time is lost, or skipped, according to Einstein's predictions. With this explanation, time is reduced by events not increased.

In looking at events I consider what an event is and find that it has minimum requirements. An event must be a change over time. Events require two different states at different times. At one moment something is one way. At another moment it is another way. Without the moments of time, it couldn't physically exist in either the first or the second way. It wouldn't be measurable. Aristotle told us that time is the measurement of change. If everything was the same at both moments, then there would be no change and no event. But if there is an event it couldn't be realized without measurement.

Existence

Physical existence is another reason why time must be fundamental, since it has a time requirement.

While we sometimes use the word "exist" to refer to something non-physical, it usually references the physical. As a result, the term has suffixes which reference time (exists, existed). These apply to our world of time and space (the physical world).

This existence is reality applied to a moment. While many things are real, they don't all exist physically, or they don't exist at the same time. Physical existence is a temporal reality. It's a reality which is specific to the world we live in, but does not have to be all of reality.

To say something exists is to say it's real now, at this moment in time. To say that it existed refers to a past reality. So, when I say "I was there a moment ago and now I'm here", I mean that I existed at that location in the past and at the present I exist at this one. Both are real, but they're not real at the same time. My existence there and here doesn't conflict because they apply to different times. And if there was no time then I could not exist physically.

It's tempting to say that time is pre-existing to all events but that's using a time reference to describe how events relate to time. So at this level of logic, I'll call it more fundamental. Time is more fundamental than "when" things were created.

Change Cannot Occur In Time

When we bring quantization into consideration, we see that change cannot occur in space or time or be recognized in any one moment. In a quantized world all events occur outside of space and time because quantization doesn't allow for transition. Change occurs between two moments, not at either of them. Those two moments make the change evident. So, there is a requirement of two moments for movement, or any change, to be realized (or made physically real).

At the foundation of science is relativity. And it tells us that the laws of nature work the same everywhere, across all three spatial dimensions and at all times, as long as your movement, even if it involves acceleration, is smooth. Tossing a bag of peanuts in an airplane works well as long as the flight's not bumpy, then it's unpredictable. The problem with this principle in our world is that all of space and time is quantized. It's not smooth. It exists in chunks. Even though they're small, you can't get from one Planck of space (the smallest measure of space) to another without stepping. And you can't get from one Planck of time (the smallest measure of time) to another without stepping. There is no in-between, no being halfway there. That's what having minimum quantities is all about. You're in one place, then you're in another. That's not movement. It's two different situations in sequence, like frames of a film.

This is the problem which led to my explanation of movement in which all objects leave space and time in order to relocate or change in any way. It does give our world a jerky quality, but Plancks of time and space are so tiny we can't notice them. One flicker between frames in a movie is a million trillion trillion trillion Plancks of time.

The inability to actually move, or change, in space or time tells me we must be leaving and coming back. Similar to how chess pieces are moved, we leave the grid of space and time then return. What I call the movement dimension is where change occurs (the "place" which is not in space or time). This is comparable to the air above the chess board. It's not part of the game. It's not part of our physical world. But without it we couldn't move the pieces or ourselves.

As movement, and all other change occurs outside of space and time, some amount of time is skipped. That amount is predicted by Lorentz's dilation formula. It is an effect of relativity which Einstein explained. And it results from movement.

Time is a dimension. And it's my understanding that dimensions are descriptors. They are not objects like matter or energy.

Descriptors are how we describe things, so they must be fixed and fundamental. We know this because the world is knowable. If descriptors could change, then our world wouldn't be knowable, and we wouldn't have reliable physical laws. Because the world is so knowable, our fundamental descriptors must be equally unchanging.

In the way that I understand time, it's a fundamental descriptor of our world. Nothing else can take its place because nothing else in our world is more fundamental. I can describe events with time, but can't describe time with events. Without the concept of time, events look like multiple simultaneous states, which is actually what the probability cloud (or wavefunction) of an object is (how scientists describe an object before it's observed, and what I say a particle arises from). This is why I suggest the wave, or probability cloud, is outside of space and time. If it weren't for time change could not be realized. An object couldn't come into existence out of its probabilities.

The physicist John Wheeler said time is nature's way of keeping everything from happening at once. This is actually not the difference between time and no time, but between only two moments of time for all to happen (or be realized) and many moments.

If it wasn't for time, nothing could happen. But if nothing happened time wouldn't cease to exist; it just wouldn't have meaning.

Now that we have things happening in a precise way, and in a precise order, let's see how they can be made to happen precisely as we want them to. The tool for our will to put new information into the physical world is a manipulation of physical laws and time, a special way of applying these structures.

Chapter 11 - Reversible Physics

Entropy is often referred to as the arrow of time, telling us which way is forward. This raises the question of what the other way is. From the present, that other direction is the past. Forward is the direction we experience time, so it's the direction we experience events. We live in forward time.

Entropy is about the arrangement of physical objects, the disordering of them. Time is separate from that. This is important because events themselves don't have a direction. That means entropy is actually about disordering or ordering.

Most laws of physics work both forward and backward, and the principle behind this is very well established in science and is apparent in some of our most fundamental theories.

Newton's third law, published in 1687 in his *Principia* (principles), tells us that for every action there is an equal and opposite reaction. This matches the first law of thermodynamics, which tells us that matter and energy cannot be created or destroyed.

Every moment must have the same amount of energy, or matter (remember that they are equivalent), as the previous one. And we identify this truth each time we write a scientific formula. That is the meaning of the equal's sign. Whatever is on the left side of the equals sign is the same or equivalent to, what is on the right. As a result, what's on the right is equal to what's on the left. And when the equation represents a process, each side of the equals sign represents a different moment in time.

Applying this to $E=mc^2$, we may see energy and matter having an equivalency. But once we look at it as an event, we see energy at one moment changing into matter at another moment. We also know that matter can change into energy.

So the fact that physics works in both directions of time is not radical at all, but instead is so fundamental that we often fail to consider it. If $a=b$ is true, then $b=a$ must also be true. *Action=reaction,* which is *cause=effect,* is also *starting energy=resulting energy*. And all these may be stated in reverse.

In our world, we're familiar with actions coming before reactions and causes before effects. This is because of the direction we experience time in relation to the event. Interestingly, Aristotle (384-322 BC) believed in "teleology" or "final cause" which means the future is as much a cause as the past is. He may have suspected that logical principles could work backwards.

In physics we may write our equations with the cause on the left side and effect on the right, or just imagine a direction to them. But the laws of physics themselves don't necessarily have this direction.

Since they seem like they do, let's look at this step by step.

Physics tells us that if one billiard ball impacts another causing the other to move, then that other could also impact the first one causing it to move. If we watched a video of these events we wouldn't know if they were played in forward or reverse time (ignoring slowing for friction).

Now let's watch one ball impact 15 balls, causing the 15 balls to move. This scenario could also be reversed, according to physics. However, as we watch it in both directions, we can see that 15 balls converging to impact one is much less probable. One ball disordering 15 balls is much more probable than 15 balls becoming ordered to move one. This is the complexity level of entropy in action.

Watching a video of an entire billiard game, we will certainly see events which appear just as likely in either time direction. And some events may appear to act against entropy. But overall, we'll easily see the move toward entropy and the true direction of time in the game, not even considering that the balls are slowing and disappearing into pockets.

The reversibility of physical laws is reasonably acceptable as we watch billiard balls interact on a table. But as soon as one falls into a pocket it becomes more difficult. This is because attraction occurs in the direction which the objects are traveling through time. To experience it as repulsion one would have to travel the other direction through time. Some objects do that.

Antimatter

The discovery of antimatter has revealed to us that for every type of particle there is also an antiparticle (protons and antiprotons, electrons and antielectrons, and so on). Two of the world's most prominent particle research centers, Fermilab located outside of Chicago, and CERN located outside of Geneva, have produced antihydrogen using antiprotons orbited by antielectrons. It seems that not just antiparticles can be made, but antiatoms too. While it's not easy, we can expect that scientists should be able to produce larger objects as well.

Something to note about antimatter is that it is not a different kind of matter from ordinary. It's made of the same substance. Antimatter is produced out of ordinary matter by reversing its properties. In this way an antiparticle actually is the particle it came from. It just has been reversed.

Antiparticles have the reverse properties of their ordinary particle counterpart. Antiprotons have a negative electric charge instead of a positive. Antielectrons have a positive charge instead of a negative. Whichever way their ordinary particle was spinning, the antiparticle spins the other way. And they're also known to be traveling backwards through time. They are so perfectly opposite that when a particle encounters its antiparticle they annihilate. Just as a square peg fitting into a square hole loses its squareness, they both disappear (or lose all their properties) leaving only energy, their fundamental component.

So if we could travel backwards in time, we would be antimatter, and we would experience gravity as pushing us. Reverse gravity would push an anti-billiard ball up out of a hole, from the perspective of the ball. Traveling backwards in time being pushed by gravity looks exactly like traveling forward in time being pulled by gravity.

What I'm suggesting here isn't a change in the direction we go through life, but a brief reversal in the direction of experience of specific particles. This would involve those particles becoming antimatter for a short time. But there's also another way to reverse time, and that is through a quantum effect.

There's a subtle difference between moving and traveling. If I walk down the street, and along the way take a step back, I've still traveled down the street one direction. This occurs at the quantum level, also. Particles travel in one direction, but sometimes move a different way. One example is heat vibration. Everything in our world vibrates according to its temperature, and these vibrations are in every direction. Meanwhile, the objects themselves have one direction of travel.

This phenomenon occurs in the time dimension, too. Just as the movement of particles through space is based on probability, so is their movement through time. While particles occasionally move backwards through space as they vibrate, they also occasionally move backwards through time. Heisenberg's uncertainty principal pairs momentum with location. It also pairs energy with time $(\Delta E\ \Delta t = h)$. This tells us that the uncertainty of time is related to the uncertainty of energy.

The inconsistency of the experience of time has been proven, as photons have been known to be absorbed before they were emitted.

All of this tells us that while we cannot travel backwards in time, time can be reversed for many events to allow the laws of nature to be executed in reverse as antimatter. And all laws of nature may be executed in reverse by the quantum effect of the uncertainty of time.

Not every physical law is reversible in forward time. As we saw, gravity is not reversible from our direction of travel through time. It can only be experienced in reverse by experiencing time in the other direction. The reason gravity is not reversible is because energy and matter are drawn together as part of their fundamental behavior. The effect we call gravity is not based on properties of objects which may be changed or canceled. Total mass or mass equivalency of energy is fixed. But other properties are changeable, such as charge, spin, and direction of experience of time.

Entropy is not an effect of something unchangeable, such as

mass. It's based on something which is changeable, the direction in which time is experienced. Entropy is based on moving through time. Because we move, we have entropy resulting from quantum uncertainty. And because particles are capable of moving the other direction through time, probabilities may also occur in that reversed time order. This makes entropy reversible. To see how this occurs we need to look for brief periods when time may have been reversed. Those periods will appear improbable in reverse time. They should stand out as appearing normal if we look at the world in reverse, similar to how some scribbles may appear as words when viewed in a mirror.

It may seem crazy to imagine our world in reverse time, but let's go down that road for a while just to see what it looks like. Let's suppose our whole world is antimatter. We see objects falling upward, cars driving backwards, broken glass assembling itself, and lottery winners handing over large amounts of money in exchange for a ticket which they will later cash in for a dollar.

None of what we see violates the laws of physics. They are the same events just with time reversed. Even exhaust fumes could flow into a car's tailpipe and converge with heat to form gasoline inside an engine, since the energy and matter resulting from gasoline burning is exactly equal to the mass of the gasoline and air it burned. This is due to the law of conservation of matter and energy. Only the lottery winner turning in his winnings is really not believable, but according to physical laws it could happen too.

Remember that in the video of the billiard game we saw events which occurred in forward time but appeared to be in reverse. A few balls occasionally acquired greater order after an event than they had before. But never do 15 balls re-assemble into a triangle as they would be at the beginning of a game. No matter how many billiard games we watch, we won't see it because it's too improbable.

Likewise, if we continue watching our world in reverse time, we'll mostly see events which are improbable. We'll see some which could happen in forward or reverse, and once in a while see one which only looks probable in reverse time. What are those? As you imagine our world in reverse time, identify which events only appear to make sense in reverse.

You'll find that some events you could identify as appearing probable in reverse time appear improbable in forward time. Those events haven't violated any physical laws. They simply only make sense in reverse time. Let's examine a couple of them closely.

That mound of sand I encountered while walking on the beach didn't accumulate by itself. A person made it. We could say that the accumulation is improbable, but appears probable in reverse time because in reverse time it appears to flatten. But that's not a full description of what happened. We can figure out that the process of a person piling up the sand is probable due to the way his body works. His body is doing what it's told and taking advantage of natural laws in the process. It's only the decision to make the pile which is unlikely. His decision, though, resulted from an accumulation of particles at a specific part of his brain, which signaled his body to act. Now look at that accumulated signal. It looks like a pile of sand. In forward time, it appears improbable that electrons would gather together. But in reverse time they appear to disperse just as the sand normally does in forward time, and that is probable.

Forming a thought appears improbable in forward time. But it appears probable in reverse time. If we form our thoughts in reverse time, then bringing them into the physical world not only doesn't violate physics, it takes advantage of it. Entropy can be reversed to bring new information into our world.

Consider the evolution of life. A more sophisticated creature is born to a less sophisticated one. Perhaps it's a creature with wings being born to one without wings. Expanding our view a little further, a creature with no wings, whose ancestors had no wings, gives birth to a creature with wings whose descendants have wings. New information is coming into the world in the form of a new design. If this is how we acquired new types of animals, this appears improbable in forward time. It really can't happen. But watch this in reverse.

Zoom in to the molecular level and look at only the brief period when the DNA and cell structure appear to make an error in duplication. You'll see that the descendant's DNA and cell structure include wings, while the parents' doesn't. It has new information in it. As new DNA and cells are built, they have an additional feature. They don't match the pattern they're made from. This can be done because in reverse time those features appear to be disintegrating or being dismantled by tiny subcellular robots. Disintegrating is entropy we're familiar with. So, in reverse time, any DNA or cell structure is probable. This is very powerful as it allows any creature to be made from another. Their only requirement, then, is that the parent can provide for the offspring as needed before and after birth.

Indications are that DNA's inability to replicate itself perfectly is due to quantum effects, the uncertainty of location creating a small chance that particles are out of place at the time replication takes place. Protons continually tunnel between the sides of the DNA strand, according to quantum probability, and could be caught on the wrong side when it separates during replication, according to researchers at Surrey University.

In forward time we see new information entering our world, a more advanced creature being born to a less advanced one. This goes against probability in the forward direction. It appears to violate natural law. But in reverse time, from a broad view, we see the probable devolution of creatures. Cells and DNA deteriorate. And in this direction, natural laws are being implemented to make it happen, particularly entropy.

I believe the quantum effect which gives us devolution can be turned around, and with careful use, give us evolution. This is because of the progression of lifeforms throughout history. Had life not followed an order of complexity through time, had primates been formed before reptiles, then we couldn't explain advancement in this way. The gradual change over time, no matter which direction it goes, is the process. The direction of time simply determines whether it appears as evolution or devolution.

Just like the higher levels of science, such as biology, emerge from the lower levels, like chemistry, the large-scale activities in our world are the result of the small scale. An atomic bomb explosion is tiny subatomic particles converting into energy. An apple falling from a tree is due to gravitons (the particles of gravity) pulling it to the earth.

An avalanche is caused by a few particles of snow or dirt falling out of place triggering others to do the same.

Quantum probability is actually not limited to small scale events. While we calculate probabilities to have a small range, those ends of the probability curve extend indefinitely. My brain doesn't need to get electrons from nearby within it. It can gather them from across the universe. When an element, such as radium, decays it doesn't have to emit radiation nearby, it could place its radiation anywhere in the universe. Nearby is just more likely.

I point out the power of probability conservatively, but there actually is no real limit to what it can do to affect our world either operating in forward or reverse time. The more dramatic events are just at the far reaches of the curve, so they're highly improbable.

Physics does allow new information into our world. Gaining new information and losing information may be the exact same process in reverse of each other.

Reversed Entropy

My proposed method for putting new information into the world is a brief and limited reversal of entropy. It's been said that the only physical law which is not reversible is entropy, as entropy is the arrow of time. I don't disagree that it generally indicates the direction we live in, but believe it's the arrow of events in time. It's the order of their execution. And this doesn't mean events can't be executed in reverse if time was reversed.

What I suggest is that events occur as

increasing entropy ▶ *decreasing entropy* ▶ *increasing entropy,*

which appear to us as

increasing entropy ▶ *new information* ▶ *increasing entropy.*

Rudolph Clausius believed that this arrow distinguished the past from the future and was the only law which could not run backwards. But that was the 1800's. Later, Paul and Tatiana Ehrenfest (friends and colleagues of Einstein) found that entropy does sometimes decrease due to quantum fluctuations. This change in view came from the discovery of quantum physics, which was after Clausius' time. The discovery of antimatter in 1932 by Carl Anderson was also well after Clausius. This doesn't mean entropy doesn't still increase in general. But it does recognize that portions of our world could experience an increase in order for a period. The acknowledged cause of this order is the same quantum probability which assures our world of overall decay.

In a broad view of our world entropy does always increase, and this is assured by quantum probability because we live in one direction. But in a narrow view of limited time and space, it has been found that those same probabilities can cause a reversal of entropy. This is the mechanism I propose for putting new information into our world, since reverse entropy is an ordering of particles in forward time. And I suggest the reason this occurs and the reason why it's not a violation is because it involves a reversal of time from the event.

But we still have 2 separate worlds. We still have a non-physical creative self and a physical self, a body made of particles.

Physical and Non-Physical

The physical world appears to be subject to laws and absolutely precise. It contains the feature of probability which prevents the future from being predictable. And reversing time allows new information to be put into the physical world by implementing physical laws in reverse.

Separate from that, we have reason to believe that there is a non-physical world and non-physical aspect to us. This non-physical aspect, not being in space or time, is not bound by any physical laws. So it may make choices, be creative, or illogical. And it may implement these decisions, or introduce this new information because for it, either direction of time is the same.

Our non-physical selves may implement the laws of physics in the direction we choose based on the outcome we want. If we agree with the predictions of nature, we go with it. If we prefer another outcome, we implement physics going in the opposite direction of time than we experience it. To create a signal for our body to act we accumulate electrons. This accumulation occurs at a future moment, then disperses in reverse time into the past.

Now we're just left with a connection issue between the two parts of us. Movement theory tells us specifically of how the non-physical connects with the physical.

Chapter 12 – Movement Theory

How does something non-physical affect something physical? My response to this is through the workings of movement theory.

Just because something which is not physical is allowed to move something physical, doesn't mean it can. If a ghost wants to pick up a candlestick, how does he grab ahold of it? Since he's not physical his hand should pass right through it in the same way that he floated through the wall.

The physical is a result of the non-physical. Everything we can touch arises from something we can't touch.

Einstein told us that matter is energy. In 1916, he wrote that, "The special theory of relativity has led to the conclusion that inert mass is nothing more or less than energy,..." In my theory of movement, I explain how it is that matter is a phenomenon of energy.

My explanation is based on the wave/particle duality of nature, which is the same as the mass/energy equivalency that Einstein told us of using his famous equation of $E=mc^2$. This equates energy (E) with mass (m) multiplied by the speed of light (c) squared. (A little mass is equal to a whole lot of energy.)

The way I explain the wave/particle (mass/energy) duality is to point out that energy is non-physical. All energy is. It doesn't even exist in space or time, yet it affects all material things. And like Einstein said, it actually is what matter is made of.

Domain

All physical things are made of particles. And those particles have no size. Instead, they have an area where they are likely to be found. Rather than space being occupied by matter, it is allotted to it similar to the way a yard is to a person. He doesn't occupy the whole yard. He moves about in it freely and dictates the rules for anyone else who may attempt to enter it.

In physics, the domain of the particle is its probability cloud. We don't know precisely where a particle is, only where it most probably is. Similarly, we don't know precisely where a person is by his address, just the lot where he could most likely be found.

Solid Objects

This may raise a concern about how it is that we seem to have solid objects. So, I refer back to my investigative method which requires that what I perceive is real. Solid objects in our world actually are solid. And this is the result of their domain.

Imagine a wall of men spaced 10 feet from each other. One might say that's not a solid wall. But tell each man that his domain stretches for 5 feet in all directions and try to walk through it. You probably won't get through, depending on the type of men (particles) they are, and the type of person (particle) you are. If you don't get through, you'll agree that it's a solid wall. If you do get through, you'll probably still agree it's a solid wall, but you broke through it.

Where my interpretation of this quantum phenomenon, the wave/particle duality, differs from most scientists is that I don't believe a person's domain goes away just because I've located him precisely. I don't think a probability cloud disappears just because a particle has been located. The probability cloud, also known as a wavefunction and visualized as a bell curve, remains.

Both the particle, which is matter because it's in the physical dimensions, and the wave, which is energy because it's in the non-physical (movement) dimension, are real together.

This can be depicted as a grid of space and time in which the wave makes an appearance at one location as a particle.

The particle will most likely appear at the high part of the curve. It may also appear at the low parts of it, which extend in all directions indefinitely, but it's not likely to. In my example of a person occupying the yard, the yard represents the high part of the curve. The neighborhood, as well as the rest of the world represents the low parts.

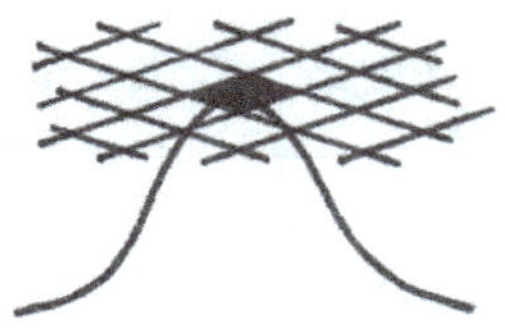

As the wave travels in the movement dimension, its presence in space and time changes location by leaving and returning just as a chess piece does. In our quantized space and time, as in chess, a particle (or piece) cannot be partially in one space and partially in another. It is in one space one moment and in another space another moment. Change occurs outside of the grid.

Quantum fluctuations (the result of uncertainty) may be imagined as that bell curve being unpredictable or wobbly such as a tall gelatin desert may be. Yet the probability curve itself moves around precisely according to physical laws.

Once we have the continued presence of the wave in a non-physical dimension along with the regular periodic appearance of the particle in space and time, resulting from the wave, we have a relationship between the physical and non-physical. We have an interface in which the non-physical is already in control of, or has a hold of, the physical. The ghost already has a hold of the candlestick.

This explanation of the physical/non-physical relationship does not necessarily include our will. It was originally intended to describe energy as a non-physical substance. The suggestion of a relationship between the energy and our will becomes a real possibility because both exist outside of space and time and therefore outside of the laws of physics. The assignment of domain for a will is not matter, it is the energy the matter is composed of, which is the probability cloud of the matter.

Proof

I've written a proof of how we know that the physical and non-physical realms do interact in this way. The details of it require an understanding of absolute relativity and appear in my book about that subject. But I'll summarize it here.

Movement theory predicts that matter which is moving is alternating between being in matter form, as a particle at rest in space and time, and being in energy form, as only a wave in the movement dimension traveling at light speed (670,000,000 mph).

We know that matter cannot pass through matter, that matter and energy interact, and energy does pass through energy. So the proof involves showing a case in which objects which we know as matter, are passing through each other.

The evidence I use comes from experiments done at particle accelerators. These are devices which cause subatomic particles to move at near light speed and then collide head-on. To accomplish this, experimenters at the accelerator I reference, create two streams of particles going opposite directions. They are each made of billions of particles of matter. So as the streams of particles are made to run into each other, there should be lots of collisions, even though some particles will miss each other just by imperfect aim. Any particles which are headed directly toward another should collide.

I predict that movement occurs by objects traveling as energy and occasionally returning to matter form to rest in space and time. If it involves leaving space and time and coming back like this, then they will pass right through each other if they both happen to be energy at the collision site. The faster the particles are traveling, the more likely they are to be in energy form and the less likely they are to be matter and able to collide. In a particle accelerator the particles move at very near light speed, so their hops outside of space and time are very large as compared to objects we're familiar with which move much slower and are almost always in matter form. The particles in an accelerator are mostly in energy form, and are rarely in matter form. Although billions of particles are aimed at each other only a few collisions occur.

My calculations show that accelerators need to use billions of particles in each stream in order to achieve the relatively few (about 20) collisions they get, because of this effect. And I believe this proves that many of the particles are passing through each other. This is predicted by movement theory and calculated using the formulas of absolute relativity. It also includes accounting for those which should miss each other due to probable distribution within the streams.

With all physical objects being made up of the non-physical, they may do what they like with the physical so long as they follow the rules of the physical world. The rules of physical behavior apply to space and time, not to the movement dimension.

Chapter 13 - Implementing Will

Through the reversal of time the improbable becomes probable. What's occurring is that we see all of the laws of nature working precisely, including the entropy which results from quantum probability, but in reverse time.

Physics is precise. While we may not have discovered all of the laws of nature, and those we do know may not be fully refined, I think we can safely conclude that they are perfect. Even quantum probability can be expected to be precise, so that all probabilities can be calculated with a precision directly proportional to the number of particles involved.

All electrons do behave exactly the same. And the same entropy which arises from probability can actually be a tool for implementing our will. As we saw in the example, the probabilistic dispersing of a concentration of electrons (just like the sand pile) is the same tool which causes the accumulation in reverse time.

One Direction

We only live in forward time. This too, is a necessary part of our lives. Living in only one direction gives our lives order and it is part of what makes our world knowable. We use the laws of nature, including entropy, to make predictions as we live.

If we lived in both directions of time we would never get anywhere. Since we would be able to go back and correct mistakes, we would. We would not know what could come in the future. In an attempt to improve our future, we would continually relive the past. There would be no progress. There would be no established history, no end. There must be completion if there is any purpose to life. Besides that, I believe there's a physical law which prevents us from living in both directions.

Conservation

The law of conservation tells us that matter and energy cannot be created or destroyed. It may only change form. Energy may become matter, and matter may become energy, but they are always equivalent. The total amount of matter/energy must stay the same. I suggest that the total amount of matter/energy must remain the same "at all times", as it travels together through time."

The total matter/energy of the universe today, according to conservation, must be the same as it was yesterday. This can be if we're all moving forward in time together. But if I were to go backwards I wouldn't be in the same moment as you are, and there would be a shortage of matter/energy in the future. If we believe that yesterday happened, then we can be sure that if I went back to yesterday there would be two of me then. I may change direction again and eventually catch up to you. And once I lived past my original moment of turnaround total matter/energy wouldn't be short by the amount of my body anymore. But yesterday would have a permanent excess of matter/energy. Conservation requires that we only live in forward time.

Earlier we saw that quantum fluctuations and antimatter might appear to be an exception to this. They are not exceptions. Antimatter does not exist twice at the same time, it's just a reversal of experience. Quantum fluctuations within time are trickier to understand.

It's true that this allows a particle to appear twice at the same moment in time as it fluctuates within its travel through time. However, while the particle appears twice, its total mass and energy only appear once.

Since mass is energy and energy is the wave which the particle emerges from, it's the wave which only has one existence. The wave is constant. Its appearance in space and time as a particle is probabilistic and therefore free to fluctuate. This includes existing, not existing, or existing as multiple particles. This actually makes time travel not absolutely impossible but practically impossible as it must occur at the far reaches of the probability curve.

Our will is not physical, which is why it can't be found in our brain. Not being physical, it's not in space or time. Our will exists outside of the physical world. It's not subject to natural laws, but it is limited. Its domain, its authority over the material world, is limited to our physical body, or perhaps just our brain. And our body is limited by physical law. In this way our will may direct our body to do anything which our body is physically capable of. Being outside of time it may participate in the world in either time direction, with both directions being bound by physical laws.

This is why we cannot see a will when it goes along with what is expected in the forward direction, when it chooses to do what's probable. We can only spot a will when it acts in a way which appears improbable in our forward way of living. But you can be sure that whatever it does, it's obeying strict physical laws, whether they be in forward time going along with the expected, or in reverse time using physical laws to make the improbable probable.

I suggest that our will can put new information (new ideas to implement) into our brain in the form of accumulations of electrons, which is the manifestation of a thought. It does this a moment ahead of where our experience is. As we experience a thought being introduced, it seems to be particles which accumulate for no scientifically explainable reason. But since our will acted in the reverse direction of time, it actually applied physical laws. It was probable.

We were given the ability to create, not by putting new matter into the world but new information, new choices.

The pairing of free will with entropy, the corruption and futility of our world, is an interesting match. At first it may seem that entropy and free will just happen to use the same tool in different directions. But they seem to be bound as necessities for each other. Perhaps decay is the cost of our freedom, or the cost of making choices. It is the quantum probabilities which cause mutations and give us physical disabilities as well as ever-changing disease. Without mutation, no child would be born with imperfections or challenges. And if there was a cold virus, we could eliminate it forever. In our decaying world, one consequence of the probabilities we will never overcome is the growing diversity of disease and disability. At least not without the help of a more powerful intervening will, as only an entity outside of the physical has the ability to stop deterioration by reversing physics.

We've been given a world which is entirely logical. And being logical, it must be precise. It's based on probability so that we may both experience it with an unknown future, and also take advantage of this probability as a tool to implement our will through reverse time. Through probability we have a perfectly logical world, and freedom in it. Someone actively participated in constructing it by forming life, then putting us in it with our own will to apply within the built-in freedom. And I believe He maintains this power to interact with the world at any place and time.

With access to the physical world which I've laid out, I think it's easy to see that the world we live in could have been created by a series of interventions. Each intervention putting new information into it to form creatures, then each of the types of life we have, and finally humanity. And while in our lives the primary source of new information we see coming into the world is human creativity, this does not eliminate the possibility of other spiritual involvement in our lives.

End Note

As I learn about physics, I find that I automatically insert two concepts of my own into my understanding of it. As any human, I bring a bias in the form of an underlying philosophy. I believe there is something beyond the physical. And I believe there's only one reality, that we all live in the same world, we just experience it differently.

I suspect many people may make the same assumptions and believe they're not even worth noting. But these concepts are not a part of theoretical physics today.

These beliefs have led me to new interpretations of physical evidence and established theories. With this philosophy I look at proven theories and try to understand how they can all make sense together. In the process I sometimes have to reduce a theory to its reliable predictions and re-imagine what they mean. This results in a new model or new way to understand it.

While trying to rationalize how Einstein's relativity and quantum physics can make sense together, I implemented a non-physical dimension. I needed something non-spatial to explain how movement is possible in a quantized world while accommodating relativistic effects.

After putting that non-physical dimension into my understanding of the world I realized I had made an assumption about relativity. I had assumed that there is only one reality which we all share. This doesn't fit with Einstein's understanding as he presented in his special and general theories of relativity. In his paper, *The Foundation of the General Theory of Relativity*, he wrote, "That this requirement of general co-variance, which takes away from space and time the last remnant of physical objectivity, ..." (General co-variance was Einstein's name for general relativity.) His explanation of all things in our world only being relative to each other, not to a common reality, is said to have eliminated absolute truth from our understanding of the physical world. But having accounted for the effects of his relativity in my theory of movement I saw that I could expand on Einstein's relativity to make more predictions than he suggests.

This process of expanding relativity led to an ability to determine earth's place in the universe.

It's in the center.

And the universe revolves around it.

By applying the concept that there's only one reality, I found a new version of relativity, an expansion of Einstein's. Applying that relativity, I found that the speed of light can be used to determine that earth is moving through space very little beyond its daily rotation. The direction of light also tells us of the same minimal movement. I further found that even inertia confirms this limited movement. This should not be surprising. Since matter is made of energy (light) it should move like it.

Applying this new relativity theory to gravity revealed that we're located at the gravitational center of the universe and that it rotates. And that rotation is what causes it to expand. A further effect of gravity which I found is the potential production of heat in bodies which are removed from the center of gravity. This heat could prevent us from venturing far from where we are. It could also prevent life like ours from forming far from here.

I can't say for sure that we're the only intelligent life in the universe, but it certainly appears that way. And I think realizing this can help us accept the great responsibility we have to our creator.

As we look up at the sky we see more stars than we can count. With telescopes we see a greater number of stars and also see more galaxies then we can count. We estimate, now, that there are about 100 billion stars in our galaxy, and about 100 billion galaxies full of stars. Many of those stars are expected to have planets around them. It's easy to suggest that with many billions of billions of planets there must be lots of life out there. However, I find that there may not be because most of those planets may be too far from the center to be suitable for life. This leads to the intimidating question of, "What are all the stars for, then, if there's no life out there and we can't visit them?"   I think the answer is simple, even from a scientific perspective.

All that happens in the universe is limited by the speed of light. This tells us that everything which happens requires a certain amount of time. Expansion, development of galaxies and stars and the various types of matter, the evolution of the universe takes time. Stars need billions of years to fuse the chemical elements we need for life.

And whatever you believe the course of the universe to be, it does have a lifespan. It began and it will end or at least become uninhabitable. In the middle of all of that change is a window for life. That window exists because the universe's life is long enough. Had the universe only lived a billion years it wouldn't have formed the chemical elements (the different types of atoms) we're made of and there couldn't be lifeforms.

We need a lot to happen for us to exist. Then we need time to live before the end. All that time, the lifespan of the universe, is determined by how much there is to do. And how much there is to do is based on how much stuff there is.

If the universe only had matter enough to make the earth and sun, its lifespan would have been less than a second, due to the speed of light. (The edge of the universe seems to move at or near light speed.) Almost nothing could happen during that time except expansion and dispersion, or collapse.

All that's in the sky, the stars and galaxies, provide the time for all that we need to happen so we can live. The stars are for us. the great number of stars in the universe provide the lifespan of the universe with a window for life in its middle. And the gravity of them provides us a central region for life. The universe is a large sparkling nest for humanity.

Many people wonder at the seeming great insignificance of themselves in this big world. But I tell you the opposite. You shouldn't find yourself small in it. You should recognize how great it is, especially since it seems to be all for you.

About the Author

Timothy Michaels is an artist and writer. Being a technically minded, "cerebral" thinker, he enjoys creating realistic art as well as exploring the world around him through physics. This led him to an explanation of color relationships using physics, which resulted in the development of his Color Calculator. Though he has no formal education in physics, Mr. Michaels has an exceptional gift for logic and an insight into the workings of the universe.

His scientific ideas rely on established theories and evidence, and are explained in a simple manner for general readership. Mr. Michaels provides new ways of understanding these theories. His ideas go beyond current science to solve problems and make new predictions. In doing so they provide the concepts and formulas necessary to advance science. Students of physics, astronomy, cosmology and other fields will find his explanations enlightening and his new ideas worth investigating.

His artwork may be viewed at: www.tmsartgallery.com

Readers are invited to comment by sending an email to: 101timsplace@gmail.com

Please put the title of the book in the Subject line.

Other Books by Timothy Michaels

Absolute Relativity: How Newton and Einstein Agree

Out of This World: The Movement Dimension

The Physics of Color Harmony

How We See Art

Walking the Cards: A Unique Drawing Method

The Universal Attraction of Gravity

Bibliography

Clegg, Brian. *The Universe Inside You*, Icon Books Ltd., 2012.

Quantum Weirdness Could Be the Driving Force Behind DNA Mutations, Science Focus, June 2022.

Baum, Eric B. *What is Thought?*, Massachusetts Institute of Technology, 2006.

Bloom, Howard. *The God Problem: How a Godless Cosmos Creates,* Prometheus Books, 2012.

Bryson, Bill. *A Short History of Nearly Everything*, Broadway Books, 2003.

Byl, John. *God and Cosmos: A Christian View of Time, Space, and the Universe*, The Banner of Truth Trust, 2001.

Carter, Rita. *Mapping the Mind*, University of California Press, 1998.

Crease, Robert P. and Goldhaber, Alfred Scharff. *The Quantum Moment: How Planck, Bohr, Einstein, and Heisenberg Taught Us to Love Uncertainty*, W.W. Norton & Company, 2014.

Darwin, Charles. *On the Origin of Species by the Means of Natural Selection,* London, John Murray, Albemarle St, 1859.

Ferris, Timothy. *Coming of Age in the Milky Way*, Anchor Books, 1988.

Fleming, Thomas A. (Editor). *Stars*, Cognella, 2011.

Fong, Peter. *Elementary Quantum Mechanics (Expanded Edition),* World Scientific Publishing Co. Pte. Ltd., 2005.

Greene, Brian. *The Elegant Universe: Superstrings, Hidden Dimensions, and the Quest for the Ultimate Theory*, Vintage Books, 1999.

Greene, Brian. *The Fabric of the Cosmos: Space, Time, and the Texture of Reality*, Vintage Books, 2004.

Greene, Brian. *Until the End of Time: Mind, Matter, and Our Search for Meaning in an Evolving Universe*, Alfred A. Knopf, 2020.

Gribbin, John, *In Search of the Big Bang*, Bantam Books, 1986.

Hall, Rupert A. *The Scientific Revolution 1500-1800: The Formation of the Modern Scientific Attitude*, Beacon Press, 1962.

Impey, Chris. *Einstein's Monsters: The Life and Times of Black Holes*, W.W. Norton & Company, Inc., 2018.

Jeanson, Nathaniel T. *Replacing Darwin: The New Origin of Species*, New Leaf Publishing Group, 2017.

Kaku, Michio. *Physics of the Future: How Science Will Shape Human Destiny and Our Daily Lives by the Year 2100*, Anchor Books, 2011.

Kaku, Michio. *Physics of the Impossible*, Anchor Books, 2008.

Kaku, Michio. *The Future of the Mind*, Anchor Books, 2014.

Lederman, Leon M. and Schramm, David N. *From Quarks to Cosmos*, Scientific American Library, 1995.

Mesler, Bill and Cleaves, H. James II. *A Brief History of Creation*, W.W. Norton & Company, Inc., 2016.

Meyer, Stephen C. *Darwin's Doubt*, HarperOne, 2013.

Penrose, Roger. *The Road to Reality*, Vintage Books, 2004.

Statham, Dominic. *Evolution: Good Science? Exposing the Ideological Nature of Darwin's Theory*, Day One Publications, 2009.

Talbot, Michael. *The Holographic Universe*, HarperCollins
     Publishers, 1991.

Wolfson, *Richard. Simply Einstein: Relativity Demystified*, W.W.
     Norton & Company, Inc., 2003.

www.ingramcontent.com/pod-product-compliance
Lightning Source LLC
Chambersburg PA
CBHW050040260726
48658CB00005B/1693